CW01486657

The future of
Antarctica

The future of
Antarctica
Exploitation versus preservation

Based on papers presented to a conference on *Antarctica: An Exploitable Resource or Too Valuable to Develop?* held at the Sir Robert Menzies Centre for Australian Studies, Institute of Commonwealth Studies, University of London

Edited by Grahame Cook

Manchester University Press
Manchester and New York
Distributed exclusively in the USA and Canada by St. Martin's Press

Copyright © Sir Robert Menzies Centre for Australian Studies 1990
Chapter 5 copyright © Commonwealth of Australia 1990

Published by Manchester University Press
Oxford Road, Manchester M13 9PL, UK
and Room 400, 175 Fifth Avenue,
New York, NY 10010, USA

Distributed exclusively in the USA and Canada
by St. Martin's Press, Inc.,
175 Fifth Avenue, New York, NY 10010, USA

British Library cataloguing in publication data
Antarctica: An Exploitable Resource or too Valuable to Develop?
    The Future of Antarctica : exploitation versus
    preservation.
    1. Antarctic. Natural resources
    I. Title   II. Cook, Grahame
    333.709989

Library of Congress cataloging in publication data
The Future of Antarctica: exploitation versus preservation/edited
by Grahame Cook.
    p.cm.
    "Based on papers presented to a conference on 'Antarctica: An Exploitable Resource
or too Valuable to Develop' held at the Sir Robert Menzies Centre for Australian
Studies, Institute of Commonwealth Studies, University of London."
    Includes index.
    ISBN 0–7190–3448–5.
    1. Antarctic regions–Congresses.   2. Antarctic regions–
International status–Congresses.   I. Cook, Grahame, 1947–
G845.F88   1990
919.8'9–dc20   90–39921

ISBN 0 7190 3448 5 *hardback*
      0 7190 3449 3 *paperback*

Phototypeset in Great Britain
by Megaron, Cardiff, Wales

Printed in Great Britain
by Biddles Limited, King's Lynn
and Norfolk
PRINTED ON RECYCLED PAPER

Contents

Foreword   *The Hon. R. J. L. Hawke, AC, MP,*
   Prime Minister of Australia   vi
Editorial preface   viii
Notes on contributors   xi
Abbreviations   xiii

1   Antarctica: an introductory overview   *David Mason*   1
2   Science as an Antarctic resource   *Richard Laws*   8
3   The mineral resource potential of Antarctica: geological
      realities   *Robert Willan, David Macdonald and*
      David Drewry   25
4   The political case for the Minerals Convention   *John Heap*   44
5   Comprehensive environmental protection of the Antarctic:
      new approaches for new times   *John Burgess*   53
6   Environmentalists' perspectives on the protection of
      Antarctica   *Kelly Rigg*   68
7   Antarctica: the legal regime   *Catherine Redgwell*   81
8   Possible future developments   *Grahame Cook*   95
    Further reading   104
    Appendices
    A   *Map of Antarctica showing territorial claims*   105
    B   *List and map of Antarctic bases*   106
    C   *Text of the Antarctic Treaty and list of Contracting Parties*   110
    D   *Text of the Antarctic Minerals Convention*   118
    Index   168

The Hon. R. J. L. Hawke, AC, MP, Prime Minister of Australia

Foreword

Debate on the issue raised by this book's title *The Future of Antarctica: Exploitation versus Preservation*, has moved swiftly in the last year. This has been due in part to the decision by France and Australia in 1989 to oppose mining in Antarctica and to work, within the Antarctic Treaty system, for a comprehensive environmental protection convention to establish Antarctica as a 'Natural Reserve – Land of Science'.

Our ideas and those of other Antarctic Treaty parties for comprehensive protection will be considered at the Special Antarctic Treaty Consultative Meeting dedicated to environmental issues in Santiago, Chile in late 1990. These developments must be seen against the new and welcome world-wide awareness of the potentially devastating effects of environmental degradation, and of the need for concerted international responses, which have prompted serious reconsideration of past policies and thinking.

The prospects for maintaining Antarctica in its pristine state will depend on the capacity of all of us to focus the attention of governments and decision-makers on the issue. Conferences such as 'Antarctica: An Exploitable Resource or Too Valuable to Develop?', and publications such as this, are an essential way of encouraging informed debate. I believe that this debate will lead to the realisation that we cannot afford to take the risks inherent in mining in Antarctica.

Antarctica is unique. This icy continent is a place of breathtaking beauty and the world's last great wilderness and is understood to play a vital role in sustaining life on this planet.

Increasingly, scientists are appreciating the role played by Antarctica in regulating the world's climate, the ocean currents and the sea level. Antarctica is important for scientific research, which itself may be vital to our own survival. It is a most valuable laboratory for measuring the greenhouse effect and changes in the thickness of the ozone layer.

I commend the Sir Robert Menzies Centre for Australian Studies, particularly Professor Millar for his work in hosting the conference, and all those who contributed to the debate on Antarctica. This is a crucial time in international consideration of the issue which has far-reaching implications for us all.

Editorial preface

Public concern about the impact of human activities on the global environment has increased markedly over the past two decades, particularly in western industrialised countries. The problems of acid rain and other manifestations of industrial pollution are only too apparent.

Discovery of 'holes' in the upper atmosphere ozone layer over the polar regions, particularly Antarctica, has drawn attention to the role of chlorofluorocarbons (CFCs) in this process and the unpredictable consequences of increased ultraviolet radiation. At the same time scientists have observed a gradual increase in the accumulation of 'greenhouse' gases such as carbon dioxide, methane, nitrous oxide, chlorofluorocarbons and ozone in the lower atmosphere. These gases threaten the global temperature balance with as yet unquantifiable implications for climate change and possible melting of the polar ice-caps resulting in a rise in sea level. The international community is working to address these global problems but environmentalists are concerned that action is not being taken far or fast enough.

Antarctica is the earth's most important heat sink and the major force driving the global climate and ocean currents. Hence any impacts on Antarctica from the greenhouse effect are likely to have global consequences. In addition, because of its extreme climate and thick ice sheets, Antarctica, alone among the continents, remains relatively unpolluted by human activities. It is, therefore, an important area of scientific research and investigation across a broad range of disciplines.

It is not surprising then, that both governments and non-government organisations have become increasingly concerned about the protection of the Antarctic environment, including its flora and fauna. Under the Antarctic Treaty system, member governments have taken action, and propose to take more, to safeguard the Antarctic environment from the impact of direct human activities. Environmentalists argue, however, that the level of protection, both in its intent and observance, is inadequate and Antarctica should be declared a World Park.

Central to the environmental protection debate is the issue of possible future minerals activities ranging from exploration through to mining. A Convention on the Regulation of Antarctic Minerals Resource Activities has been negotiated under the Antarctic Treaty system on the assumption that minerals activities may be possible at some time in the future without inflicting undue environmental harm.

Since the Minerals Convention was adopted by the Antarctic Treaty parties on 2 June 1988, several countries, particularly France and Australia, have had second thoughts about proceeding to sign and/or ratify it. Instead of proceeding with signature of the Minerals Convention, France and Australia have proposed that a comprehensive environmental protection regime be negotiated which would include a ban on minerals activities in Antarctica. In contrast, other Treaty parties believe that the Minerals Convention provides a mechanism for enhanced environmental protection while leaving open the question as to whether or not minerals activities could be responsibly and safely undertaken at some future time. The United Kingdom, for example, has already passed legislation which will enable it to ratify the Minerals Convention.

The decision of France and Australia has effectively prevented the Minerals Convention from coming into force. This has resulted in an element of tension between the Antarctic Treaty parties. All the Antarctic Treaty parties have agreed, however, that it is desirable to implement a more comprehensive regime of Antarctic environmental protection. A special meeting of the Antarctic Treaty Consultative Parties has been agreed for late 1990 to discuss this issue. In addition, a conference to review the Antarctic Treaty itself may or may not be held in 1991 when the initial thirty-year period of the Treaty expires.

It was against this background that the Sir Robert Menzies Centre for Australian Studies decided that it could make a contribution to better understanding of the issues by organising a conference to provide an opportunity, outside the usual 'Antarctic club' gatherings, to debate the arguments for and against the Minerals Convention and the natural reserve/world park concept. To that end we thought it

important to obtain the participation of a broad range of interested
parties with and without a particular point of view to express. The
response to the Centre's conference, 'Antarctica: An Exploitable
Resource or Too Valuable to Develop?', exceeded our expectations.

The publication of this volume represents the outcome of the
conference. It begins with an overview of the development of the
Antarctic Treaty system and the background to the present debate
about the Minerals Convention. Next the environmental and scientific
importance of Antarctica is explained and what little is known about
its potential minerals resources. Three chapters are devoted to
examining the political and environmental arguments for and against
the Minerals Convention, including the environmentalists' perspect-
ive. Some of the legal issues are then explored and the concluding
chapter looks at possible future developments. The appendices
provide useful background documents and information.

This comprehensive and multidisciplinary approach provides an
excellent overview of the debate about appropriate future environmental protection of Antarctica, particularly in relation to possible
exploitation of its little known mineral resources. We have
chosen to present the material in a simple style suitable for both the
academic and general reader interested in Antarctica or environmental
issues generally.

The Sir Robert Menzies Centre for Australian Studies wishes to
record its appreciation to the various contributors to this book, to
Robert Headland for providing the maps of Antarctica, to David
Mason for his valuable participation in the Steering Committee for the
Conference, and to the staff of the Centre who did so much to make
this project a reality. In particular we wish to thank Wendy Robins for
preparing the manuscript and Kirsten McIntyre for her administrative
assistance.

Grahame Cook
Sir Robert Menzies Centre for Australian Studies
April 1990

Notes on contributors

Mr David Mason holds a Master of International Law degree from the Australian National University, and is a former Head of the Antarctic Section of the Australian Department of Foreign Affairs. He is at present Counsellor (Political) at the Australian High Commission in London.

Dr Richard Laws CBE, FRS, FI Biol., is a former Director of the British Antarctic Survey and is currently Master of St Edmund's College, Cambridge. He is a leading Antarctic scientist and has written extensively on Antarctica, including *Antarctica: The Last Frontier*, published with Anglia Television.

Dr Robert Willan has worked as a geologist in Africa, Europe, Australia and the USA. He joined the British Antarctic Survey as Head of the Mineralisation Geology Section in 1985 and has had three field seasons in Antarctica.

Dr David Macdonald worked for the British Antarctic Survey from 1975 to 1980 on South Georgia. He gained experience of the oil industry with British Petroleum and rejoined the British Antarctic Survey as Senior Sedimentologist in 1984.

Dr David Drewry is Director of the British Antarctic Survey, Chairman of the Council of Managers of National Antarctic Programmes, and Vice-President of Comite Arctique International. He is a leading Antarctic scientist and has published extensively on Antarctica.

Dr John Heap is Head of the Polar Regions Section of the United Kingdom Foreign and Commonwealth Office. He has had extensive experience in Antarctic matters, representing the United Kingdom at numerous high level conferences. He is editor of the *Handbook of the Antarctic Treaty System*.

Mr John Burgess is Assistant Secretary, Environment and Antarctic Branch of the Australian Department of Foreign Affairs and Trade. He is a former Australian Ambassador to Poland and has represented Australia at international conferences on Antarctica.

Ms Kelly Rigg is Antarctica Project Director for Greenpeace International. She has been working on the World Park campaign for the last seven years and oversees all aspects of Greenpeace's Antarctic effort, including its annual expedition to Antarctica.

Ms Catherine Redgwell is a lecturer in law at the University of Manchester. She is qualified as a barrister and solicitor in the Province of British Columbia, Canada, and holds a Masters degree in Sea-Use Law, Economics and Policy from the London School of Economics and Political Science.

Mr Grahame Cook is at present Australian Public Service Fellow at the Sir Robert Menzies Centre for Australian Studies. He holds an economics degree from the Australian National University and is a Senior Adviser in the Australian Department of the Prime Minister and Cabinet.

Abbreviations

AEIMEE   Group of Specialists on Antarctic Environmental Implications of Possible Mineral Exploration and Exploitation

AMCAFF   Agreed Measures for the Conservation of Antarctic Fauna and Flora

ASOC   Antarctic and Southern Ocean Coalition, comprising Greenpeace and more than 200 other members in thirty-five countries, with secretariat offices in Washington, DC, Wellington and Sydney

ATCM/WP   Antarctic Treaty Consultative Meeting/Working Paper

ATCM   Antarctic Treaty Consultative Meeting

ATCP   Antarctic Treaty Consultative Party

ATP   Antarctic Treaty Parties

ATS   Antarctic Treaty system

BAS   British Antarctic Survey. One of the research institutes of the Natural Environment Research Council

BIF   Banded iron formations

BIOMASS   Biological Investigations of Marine Antarctic Systems and Stocks

BOE   Barrel of oil or oil equivalent

CCAMLR   Convention on the Conservation of Antarctic Marine Living Resources

CCAS   Convention for the Conservation of Antarctic Seals

CFCs   Chlorofluorocarbons, a family of inert, relatively non-toxic gases mainly used as propellants in spray cans, refrigerant gases, solvent cleaners and for foam plastics

CRAMRA   Convention on the Regulation of Antarctic Mineral Resource Activities

DDT   A chlorinated hydrocarbon insecticide, Dichloro-
            diphenyl trichlorethane, widely used between the early
            1940s and 1960 when stringent restrictions on its use
            began to be imposed

EAMREA   Group of Specialists on Environmental Impact As-
            sessment of Mineral Resource Exploration and Ex-
            ploitation in Antarctica (predecessor of AEIMEE)

ICSU   International Council of Scientific Unions

IGBP   International Geosphere Biosphere Programme

NERC   Natural Environment Research Council

SCAR   Scientific Committee on Antarctic Research. A scientific
            committee of the International Council of Scientific
            Unions

UK   United Kingdom

USA   United States of America

USSR   Union of Soviet Socialist Republics

1 *David Mason*

Antarctica: an introductory overview

The debate about the future of Antarctica, particularly whether, like other continents, it should be viewed as an exploitable resource, or rather as simply so important environmentally as to rule out development activities, invites consideration of man's impact so far on the southernmost continent and its surrounds. Man's involvement in Antarctica has proceeded through four phases.

First, as recently as 1773, there appeared virtually no prospect that Antarctica would be discovered, let alone explored. Even as intrepid a navigator as Captain James Cook, the first person known to cross beyond the Antarctic circle to a longitude of just over seventy-one degrees, wrote in his journal: 'I can be bold enough to say that no man will ever venture further than I have done and that the land which may lie to the south will never be explored.'[1] But as we know, some fifty years later, that land to the south – the Antarctic continent – was discovered and, although sealers had been active in the Antarctic region since the 1780s, the first recorded landing was made in 1821.[2] The process of exploration and exploitation had thus begun.

During the following century, adventurous men made their mark in Antarctica, albeit at great cost to their health if not their lives. The concept of international co-operation in Antarctic scientific research was accepted relatively early and resulted in the first International Polar Year in 1882–3. Following the Sixth International Geographic Congress in London in 1895 there was enormous international interest in Antarctic exploration. However the South Pole was not reached until late 1911. In this second phase, Antarctica was perceived as 'the pure and dreadful continent' to quote the Australian novelist, Thomas Kenneally.[3]

As time went on, the dread in which the Antarctic continent was held abated somewhat, increasingly to be replaced by a desire to further explore Antarctica's mysteries. At first, the subject of interest was an intangible one – knowledge about the continent – but national interests also came to the fore. Between 1908 and 1940 seven nations made territorial claims to sovereignty over various sections of Antarctica (see Appendix A). Permanent scientific research stations were established in the Antarctic from the early 1940s (present bases are listed in Appendix B). There also quickly developed great interest in exploiting the extraordinarily abundant living resources – krill, fish, seals and whales – found in the waters around Antarctica. Antarctic whaling commenced in the early 1900s and became a major industry before over-exploitation lead to the implementation of controls through the International Whaling Commission to protect threatened species. More recently there has been increasing speculation about the possibility of exploiting whatever mineral resources may exist within the Antarctic land mass and its continental shelf.

From the outset of this third phase, in which Antarctica may be deemed to have assumed scientific and economic value, involvement by governments and international co-operation became hallmarks of man's activity in the region. A second International Polar Year of scientific activity was held in 1932–3. The interest generated in further co-operative scientific research led ultimately to the International Geophysical Year of 1957–8.

Resource issues

The success of the international scientific co-operation carried out during the International Geophysical Year, particularly in Antarctica, provided the impetus for negotiation of the 1959 Antarctic Treaty (*see* Appendix C). The key concerns of the Treaty are to ensure that Antarctica is used exclusively for peaceful purposes; to prohibit nuclear explosions and disposal of nuclear waste; to guarantee the freedom of research and ensure scientific information exchange about Antarctica, and to establish a system of on-site inspection to promote the objectives of the Treaty. It is open to any state to become a party to the Treaty and they may obtain Consultative Party status by committing themselves to a programme of Antarctic scientific research.

It is worth noting that neither living nor non-living Antarctic resources receive more than scant attention in the Antarctic Treaty itself. Rather, Antarctic Treaty Consultative Parties (ATCP) have attempted to deal with these tangible resources and to regulate their exploitation through development of a number of agreed measures, accepted practices and three separate conventions.

The first of the conventions, the Convention for the Conservation of Antarctic Seals (CCAS), was concluded in London in 1972. It extends to seals located in the Antarctic pack-ice some of the protection provided to seals found on or near the Antarctic coastline by the Agreed Measures for the Conservation of Antarctic Fauna and Flora (AMCAFF) negotiated under the Antarctic Treaty in 1964. The Convention has been successful in preventing a recurrence of the decimation of some species of Antarctic seals.

More significant, however, was the later Convention on the Conservation of Antarctic Marine Living Resources (CCAMLR), concluded in Canberra in 1980. This deals with Antarctica's abundant fishing resources and seeks to manage and conserve the marine ecosystem in the waters south of the Antarctic Convergence. Adopting an ecosystem-wide approach it stipulates, *inter alia*, that no harvested species should be allowed to fall below a level close to that which ensures the 'greatest net annual increment' and that the ecological relationship between harvested, dependent and related populations should be maintained. It is worth recording, however, that at the November 1989 Hobart meeting of the Commission for the Conservation of Marine Living Resources, the policy-making body established under the Convention, some parties expressed their conviction that these ambitious aims were not being met.

The 1988 Minerals Convention

The third resource convention – and unlike the first two, this has not entered into force – is the Convention on the Regulation of Antarctic Mineral Resource Activities (CRAMRA – *see* Appendix D). This was adopted by a Special Antarctic Treaty Consultative Meeting in Wellington in June 1988, after six years of negotiations which had their origins in consideration of the question during the 1970s. CRAMRA aims to provide a framework in which decisions may be made to permit exploration and mining on the Antarctic mainland and continental shelf should mineral exploitation ever become technologically and commercially feasible. It stipulates that no activity can occur unless there is adequate information to judge its environmental impact, and unless the activity in question is judged not likely to affect adversely the Antarctic environment. In the case of prospecting activity, the necessary judgements have to be made by the State authorising it, although they may be challenged by other members of the CRAMRA Commission. No exploration and development activity can occur until there has been agreement by all members of the Commission to open an area for such activities.

The Convention on the Regulation of Antarctic Mineral Resource Activities breaks new ground insofar as it purports to put the onus on those wishing to undertake mineral resource activities to prove that these could be carried out in accordance with the environmental standards set out in the Convention. This contrasts with the Antarctic Marine Living Resources Convention (CCAMLR), where commercial fishing is permitted without there being a need to demonstrate in advance that such activity would not threaten fish stocks.

Environmental concerns

The fourth phase, or perhaps it is really an overlay on the third phase, of man's activities in Antarctica is characterised by an emphasis on conservation and the protection of certain aspects of the Antarctic environment. It would be wrong to suggest though, that the Antarctic Treaty Parties have only recently become aware of Antarctica's environmental significance. On the contrary, the record of the Consultative Parties on environmental protection has been an impressive one.

The Antarctic Treaty, the three resource conventions mentioned earlier, and the 1964 Agreed Measures for the Conservation of Antarctic Fauna and Flora all bear testimony to this. In 1970, the item entitled 'Man's impact on the Antarctic environment' made its appearance on the Antarctic Treaty Consultative Parties' agenda, leading to an additional range of measures. This history and the current impetus to extend the scheme of protected areas in Antarctica reflect a widening perception of what is worthy, and in need of, protection. Moreover, at the October 1989 Antarctic Treaty Consultative Meeting in Paris, a number of important recommendations were adopted on environmental issues, for example, on waste disposal and marine pollution.

But it is a measure of how much things have changed that the Antarctic Treaty itself, when drafted in 1959, contained no specific reference to the preservation of the environment in the guiding principles set out in its preamble or in its key substantive articles. Indeed, the word 'environment' is not used at all in the Treaty; the only environment reference is to the responsibility mentioned in Article IX for the 'preservation and conservation of living resources in Antarctica'.

Official reaction has thus moved a long way towards according significant environmental protection in response to scientific, resource-based and general environmental concerns. There is, however, a need for the Antarctic Treaty parties not to stand on this record but to anticipate and prepare for activities or developments which affect the environment. Indeed a strong case can be made that Treaty Parties have not yet made sufficient response to the recent rise in international concern about the Antarctic environment, and the increased awareness of the key interrelationships between Antarctica and the global environment.

It has, of course, long been recognised that Antarctica plays a crucial role in global climate matters, influencing our weather, the ocean currents and the sea level. Antarctica's role as the world's most valuable laboratory for measuring the greenhouse effect and changes in the thickness of the ozone layer are also well known. But only recently has there developed a widespread view that Antarctica is perhaps not the invulnerable, inhospitable environment impervious to all change it was once thought to be, but rather a fragile ecosystem whose preservation is in some way linked to the future health of earth itself.

It is certainly the case that through various media, including some spectacular television programmes produced in Britain,[4] the general public in numerous countries has been made acutely aware of these issues, to the point where Antarctica is no longer a specialist subject primarily of interest to scientists and resource experts, but one of fascination and concern to laymen everywhere.

It can be strongly argued that those countries, charged under the Antarctic Treaty system with ensuring that Antarctica be treated in the interests of all mankind, cannot be unaffected by the dramatic rise in international concern for the environment. Australian concerns on this point focus particularly, although by no means exclusively, on the mineral exploitation aspect. Australia argues that the factors identified above, including the qualitative change which it believes has occurred in international concern about environmental issues, challenge the assumption made when the mineral negotiations began in the 1970s, that mineral resource activity could theoretically take place consistent with adequate protection of the Antarctic environment. Australia has thus expressed strong doubts as to whether in the foreseeable future mining in Antarctica could ever be safe. It has also voiced concern that CRAMRA could become an incentive to mining, since it permits under certain circumstances mineral prospecting without institutional authorisation, and sets out a procedure on exploration and development, which if met by a mining operator would ultimately allow that operator to obtain security of title and thus be encouraged to conduct mining operations.

1989 Comprehensive environment protection convention proposals

In a television interview on 20 April 1989, the French Prime Minister, M. Rocard, called for a re-opening of Antarctic minerals negotiations to take better account of environmental concerns.[5] At the same time the Australian government was carefully considering whether to sign the Minerals Convention. On 22 May 1989 Australia announced that it is 'dedicated to the comprehensive protection of the Antarctic environment and in that context our strong commitment is that no mining at all – including oil drilling – should take place in and around the continent'.[6] It added that 'Australia will not sign the Minerals Convention, but instead will pursue the urgent negotiation of a comprehensive environmental protection convention'. Later Australia announced its development with France of a proposal for establishing Antarctica as a 'natural reserve, land of science'.[7]

In essence the Australian–French proposal is for a new legal instrument negotiated under the Antarctic Treaty umbrella which would seek to provide a set of principles of environmental protection applicable to all human activity in Antarctica and procedures and machinery for implementing them. These would permit rapid and ongoing responses to environmental problems. Moreover, it would provide a system which would anticipate those problems which may arise, as well as providing for monitoring and surveillance on an ongoing basis.

Conclusion

Unlike CRAMRA, the details of which are set out for all to see, the various elements of the comprehensive Antarctic environmental protection convention proposal put by Australia and France remain to be fully developed and elaborated. Yet it is already clear that what is envisaged marks a new approach from CRAMRA, not only on the particular point of Antarctic mineral resource activity, but also in its departure from the sectoral or 'bottom-up' approach traditionally favoured by the Antarctic Treaty system. These differing approaches make a comparison of the merits of the two a worthy and stimulating focus for discussion. The central issues in the present debate are examined in the succeeding chapters.

Notes

1 J. C. Beaglehole (ed.), *Journals of Captain James Cook on His Voyages of Discovery: The Voyage of the* Resolution *and* Adventure *1772–1775*, Cambridge University Press, 1961, p. 638.

2 For a general overview of man's activities in the Antarctic see R. M. Laws, *Antarctica: The Last Frontier*, Boxtree, London, 1989, pp. 155–204.

3 T. Kenneally, *The Survivor*, Penguin Books, Ringwood, Australia, 1970, p. 10.

4 For example, the Anglia Television programme, *Antarctica: The Last Frontier*, produced by Graham Creelman.

5 Reported in *Le Monde*, 24 April 1989, p. 4.

6 R. J. L. Hawke, Prime Minister of Australia, *Protection of the Antarctic Environment*, media statement, Canberra, 22 May 1989.

7 See *Joint Statement on International Environment Issues Agreed by Prime Ministers Hawke and Rocard*, media statement, Canberra, 18 August 1989; and R. J. L. Hawke, *Antarctica's Future: Continuity or Change?* Speech to the Sixteenth National Conference of the Australian Institute of International Affairs, Hobart, 18 November 1989.

Science as an Antarctic resource

The Antarctic environment

The Antarctic comprises nearly a tenth of the earth's surface, that is, about a tenth of the world's land surface and a tenth of the oceans. For this reason alone it is important to improve our knowledge about the Antarctic region. As the highest, windiest, coldest, and driest continent, it has the most rigorous environment on earth. It is also the most remote region, without an indigenous human population and so most nearly approaches a pristine condition.

In order to understand Antarctica's value to science we first need to understand its geography, the nature of the environments that characterise it, the geophysical patterns and processes, the broad geology and the life forms of land and sea. The purpose of this chapter is to give only a brief overview but fuller accounts can be found elsewhere, for example, in *Antarctica: The Last Frontier*.[1]

The coastal fringes and mountain tops now constitute the mere two per cent of the continent that is not covered by ice (up to four kilometres deep in places). Although during periods of warmer climate in the remote past trees and vertebrate animals occurred, because of extreme cold and ice, life on land is now largely limited to microbes, lichens and mosses and to inconspicuous invertebrate animals.[2] More than forty per cent of the Antarctic coastline is fringed with ice shelves or glaciers and it is the intervening ice-free stretches that provide the most favourable environmental conditions for plants and invertebrate animals, as well as coastal breeding sites for seabirds and some seal species. These limited snowfree coastal areas also provide the best sites for research stations, so it is here that the most obvious environmental impacts are focused. In the cold, relatively dry conditions, soils form very slowly and the fragile land and inland water ecosystems are very susceptible to disturbance and pollution.[3]

The marine environment is more stable, its ecosystems more diverse and robust. The seas contain sustainable commercial resources of krill, fish and squid, possibly exceeding total annual world fish catches which have peaked at about eighty million tonnes. The history of earlier over-hunting, first of seals in the nineteenth century and more recently of whales, is well known, and has had enormous consequences for the other components of the Southern Ocean ecosystems. These perturbations have unintentionally provided biological science with a vast ecological experiment, an important case study which helps us to understand the dynamics of ecological interactions elsewhere.[4] In global terms, however, the current commercial catches of Antarctic living resources are relatively insignificant. Because of the buffering powers of the vast Southern Ocean with its enormous current systems, the seas are also less vulnerable to other indirect environmental impacts from human activities than the land.[5]

Antarctic mineral resources have not yet been the target of commercial exploitation and there are no proven commercial resources, either of hard rock minerals or of hydrocarbons.[6] Their actual occurrence is speculative, and it seems likely that full-scale commercial exploration and extraction will only take place when the more accessible reserves elsewhere are seriously depleted or their supply is affected by political change. In other words, there would have to be a transformation of the relevant economic conditions because Antarctic operations are very costly.[7] For logistic reasons mineral extraction if it occurs will probably have the greatest environmental impacts on those accessible coastal regions mentioned earlier; however, dust produced by mining operations could have a wider effect on the ice sheet.[8] Hydrocarbon extraction if it occurs would most likely exert its major environmental effects at sea.

The overall importance of the Antarctic to mankind, however, is probably much greater, more immediate and quite different from that suggested by commercial scenarios. The apparently untouched ice-cap itself constitutes an archive of man-made pollutants, the rate of deposition of which reflects changes in world-wide baseline levels.[9] More fundamental for mankind are the implications of Antarctica for the global environment. It is the earth's major heat sink, because of its extensive ice blanket which reflects incoming solar radiation, resulting in an extreme temperature gradient separating it from lower latitudes. Consequently it exerts a major influence on the world's climate and drives major ocean currents which influence the world's oceans, so that changes in the Antarctic are likely to have global impacts.[10] The region is likely to be affected by future climatic warming with consequential effects on sea level and the protective overlying ozone layer has already been drastically thinned.[11]

Despite this the Antarctic probably remains more natural and unaffected by human activities than any other region. Antarctica's size and wider influence means that it cannot be ignored in a world which now realises that global changes affect all peoples. This, together with the contribution to knowledge of other Antarctic sciences, leads many of those involved to believe that the region's main resource is its potential for scientific discoveries that help us to understand conditions and processes elsewhere, as well as its contribution to understanding and monitoring global change.

The importance of Antarctic science to the wider world can be demonstrated by examining its past achievements and future potential, and by describing some of the problems to be solved and opportunities to be grasped. This chapter only provides a broad summary of some of the highlights by means of examples. Fuller accounts can be found in many of the publications cited earlier and in *International Research in the Antarctic*.[12] Such research is planned and co-ordinated by the Scientific Committee for Antarctic Research (SCAR) of the International Council of Scientific Unions (ICSU).

Meteorology and climatology

The fragmentation of the supercontinent of Gondwana and the breaking of the connections between the Antarctic continent and South America and Australia, made possible the formation of the Circumpolar Current some twenty-seven million years ago.[13] This, the largest current system in the oceans, links the Atlantic, Indian and Pacific Oceans but isolates the Antarctic continent from the influence of their warmer waters.[14] It is the main factor promoting and sustaining the formation of the high cold dome of the Antarctic ice sheet and the surrounding belt of pack-ice, which waxing and waning seasonally, effectively doubles the size of the continent in winter. Incoming radiation from the sun is of lower intensity at higher latitudes and the land and sea ice are strong reflectors, so the Antarctic has a large radiation deficit which has a major effect on the global circulation.[15]

In addition, the cold Antarctic Bottom Water which forms over the continental shelves below the ice shelves, flows northwards and influences the oceans and weather far beyond the Antarctic, even into the North Atlantic.[16] Global atmospheric modelling and regional weather forecasting, therefore, depend increasingly on more detailed data from the polar regions. The Antarctic is also a natural laboratory for studies of atmospheric processes like the boundary layer that forms as cold air flows over an ice shelf. Research on such aspects of atmospheric dynamics can contribute to an understanding of similar processes elsewhere on earth.[17]

From the autumn onwards in the Antarctic there is a build-up of strong thermal gradients between the increasingly colder high latitudes and warmer low latitudes. The ring of westerly winds strengthens and, like the circumpolar current in the ocean, isolates the atmosphere over the continent from the regions to the north so that a stable polar vortex forms with a cold, dense, relatively still core of air centred on the pole. The vortex persists until it breaks down in the spring.[18]

Attention has recently been focused on two crucial changes in the atmosphere; the thinning of the ozone layer and the related atmospheric warming due to the 'greenhouse effect'.[19]

The recent discovery of the thinning of the ozone layer over the Antarctic in spring by the British Antarctic Survey is the prime example to date of the value to the world of research in the Antarctic.[20] The ozone layer filters out lethal ultraviolet radiation from the sun and the present life on earth depends as much on this filter as on oxygen or water. The ozone hole extends over about four per cent of the earth's surface in spring, and a similar but much less pronounced condition develops over the Arctic. It is now known that the Antarctic vortex provides conditions in the still and extremely cold central core for the formation of polar stratospheric clouds at heights of twelve to twenty kilometres. Within these clouds ice particles act as nuclei on which a multitude of very complex chemical reactions take place, culminating in the destruction of ozone at the time of the spring warming by the returning sun. These reactions are known to occur nowhere else on earth and balloon and aircraft flights into the ozone hole in spring have now sampled conditions at these altitudes. Measurements of ozone, aerosols, water and various chemicals established beyond doubt the role of man-made chlorofluorocarbons (CFCs) in destroying ozone. The chemical reactions are still poorly understood, but the chlorine cycle is probably the most significant.[21]

The thinning of the ozone layer, most obvious over the Antarctic, has extremely serious implications for all kinds of living organisms, although ultraviolet radiation has limited penetration into water. If this academic research and monitoring had not been initiated in 1957 and continued for over thirty years the world would still be in blissful ignorance of a major threat to life on earth and no steps would have been taken internationally to phase out the production of CFCs.

Global warming, which appears to have begun, will be greater nearer the poles and the Antarctic is crucially important in this context.[22] The initial process is that incoming radiation from the sun warms the ground which then generates long wave infra-red radiation. Clouds and gases in the atmosphere (notably water vapour, carbon dioxide, methane, ozone, nitrous oxide and CFCs) absorb the infra-red radiation and heat up; as a result the temperature of the whole atmosphere rises and, in turn, that of the oceans. The so-called greenhouse gases have been increasing, largely due to human activities. The Antarctic ice sheet has preserved within it an archive of changes in climate[23] and of the greenhouse gases that can be studied by means of bubbles in ice cores which represent atmospheric composition and temperature at the time they became trapped. The longest of these has enabled scientists to look back 150,000 years.[24] Such cores also record the changes over time in pollutant levels in the atmosphere – heavy metals, insecticides, CFCs and others.

Carbon dioxide is still the most important greenhouse gas and atmospheric concentrations of this gas have increased by about twenty-five per cent since 1850, mainly due to the burning of fossil fuels and the burning of the world's forests which has released carbon previously locked up in organic compounds.

The rate of increase would be greater but for the oceans whose biological and physical processes remove carbon dioxide from the atmosphere. The Southern Ocean alone removes in this way about thirty per cent of carbon dioxide produced; damage to its plankton communities due to pollution, ultraviolet radiation or warming could therefore have serious consequences. Research in the Antarctic is vital to our understanding of the processes involved.

As they warm the oceans will expand, causing the sea level to rise world-wide. This could lead to the breakup of the West Antarctic ice sheet, much of which is grounded below sea level. At the same time it is possible that melting of the ice sheet and contraction of the pack-ice might be brought about by higher air temperatures. If the ice sheet melted completely, which is probably only a remote possibility, it is estimated that there would be a fifty-five metre rise in world sea levels.[25] Although none of this is certain and other scenarios have been advanced, a rise of one-half to one metre is possible within a century. Even such a minor change would have serious consequences for low-lying countries and cities at sea level would have major problems. Research is needed to establish the mechanisms which control the growth, flow and decay of the ice sheet, in order to construct predictive mathematical models. Here, satellite altimetry to describe the surface of the ice sheet and radio-echo sounding to establish its base can provide estimates of the total volume of ice and show how it may be changing.[26]

Geospace

About ninety-nine per cent of all matter in the universe is in the form of plasma (an extremely thin electrically charged gas) and the upper atmosphere, above about seventy kilometres, is also a plasma – the ionosphere – formed under the influence of X-radiation and ultraviolet radiation from the sun. The magnetosphere is a volume of space near the earth that is dominated by the magnetic field lines connecting the northern and southern polar regions, which determine the behaviour of the plasma and so shield the earth from the solar wind. The ionosphere and magnetosphere, taken together are known as Geospace.[27]

The solar wind is a continuous supersonic stream of electrically charged particles from the sun, which gusts as events occur on the sun. These particles flow towards the earth and its ionosphere along the magnetic field lines and interact to give the conspicuous auroral displays and disturbances. This allows deep-space phenomena to be studied from the ground and the Antarctic is very well placed for such studies.[28] For example, scintillation of celestial radio sources makes possible analysis and prediction of disturbances which are followed by geomagnetic storms. Ionospheric research, in addition to its basic importance for the understanding of processes in geospace, also has practical application in relation to radio communications.

The magnetospheric shield, like the ozone layer, is crucial to life on earth.[29] Also, magnetic storms endanger satellites and astronauts out in space. Closer to earth, concern is currently being expressed about the exposure to these particles of people in high flying commercial aircraft over the Arctic. This exposure will increase with later generations of aircraft or rocket transport vehicles and so prediction of magnetic storms is essential. Although the Antarctic polar air route is not yet in routine use to connect the southern continents, the situation is likely to change with technological developments as the projected increase in Southern Hemisphere populations occurs and their standards of living change.

Geology and geophysics

Serious research has been under way for a little over a quarter of a century, handicapped by the limited rock outcrops, the remote and hostile environment and logistic problems. Knowledge of Antarctica is very limited compared to the rest of the world. The exposure of only two per cent of the rocks means that remote sensing geophysical techniques must be used to investigate the structures in the large and deep subglacial areas. The land-based studies have to be complemented by geophysical and geological research on the ocean basins and continental shelves; this is complicated by the difficulty of working in the pack-ice zone and the risk of losing expensive equipment in ice-infested seas. The broad picture is established but much detailed work is still necessary.[30]

A major concern of earth scientists is the breakup of the original supercontinent of Gondwana, by continental drift, studied in the light of plate tectonics theory.[31] This is a problem of global significance because the Antarctic plate is contiguous with seven other lithospheric plates and such studies lead to a better understanding of major processes in the earth's mantle. One method is to make reconstructions of the situations at different times in the distant past; the major uncertainties are about the Pacific margin from South America to New Zealand.[32]

Within the present Antarctic continent Lesser Antarctica is one of the important regions for the study of tectonics, for it is composed of a number of micro-continental blocks, the largest of which is the Antarctic Peninsula.[33] There is interest in the nature of these blocks, which have moved relative to each other and to the massive continental shield of Greater Antarctica; and in the intervening subsea and subglacial basins which are floored by sedimentary structures that may have potential as hydrocarbon reservoirs.[34] Mineral occurrences have been recorded and this work will be useful if and when full-scale mineral exploration begins.

The Antarctic Peninsula and the Scotia Sea are among the best regions on earth in which to study large-scale basic geological and geophysical processes.[35] The western margin of the Antarctic Peninsula provides one of the best examples of a spreading centre migrating into a subduction zone, and helps to determine the rates of subduction processes. It also contributes to the interpretation of similar processes occurring in more complicated regions. In addition, associated mineralisation processes and metallogenesis have resource implications and parallels can be drawn with South America. This magmatic arc has been active for some 180 million years which makes it one of the longest lived; it is unusual because fore-arc and back-arc terrains are well exposed. These two factors make it one of the best regions on earth for studying the interactions between tectonics and sedimentation, so results from investigations there have a global significance. Other exciting Antarctic work in this field relates to the opening of the Weddell Sea more than 160 to 170 million years ago. Upper Mesozoic and Cenozoic strata are preserved in the bottom sediments of the Weddell, Ross and Amundsen Seas, which represent probably the largest unexplored body of sediment in the world that remains to be assessed for hydrocarbon resource potential.[36]

Palaeomagnetic studies have shown that parts of Gondwana moved through different latitudes and have experienced great climatic changes.[37] Palaeobotanical investigations, including detailed tree ring studies, together with other investigations, such as coring of sea floor sediments and the ice sheet, also contribute to the reconstruction of climatic history. The fossil record is fragmentary owing to the limited amounts of sedimentary rock exposed, but the distribution of fossils during different periods can be used in biogeographical studies involving the fauna of all the southern continents. Marine fossil faunas suggest that the region between Greater and Lesser Antarctica was an important centre of origin for evolutionary radiations which then spread northwards into warmer waters. On land the presence of Triassic *Lystrosaurus* fossils in the Transantarctic Mountains demonstrates the faunal links between the present southern continents. Fossil marsupials on Seymour Island help understanding of the dispersal of land vertebrates and indicate that there was a land bridge with South America into the Cenozoic period, when the Drake Passage opened. There are many other examples of fossil records.[38]

Biology – the Southern Ocean

The accepted boundary of the Antarctic is the Antarctic Convergence or Polar Front, a physical boundary found on average at fifty degrees South latitude, where cold, northward-flowing Antarctic surface water meets warmer southward-flowing water.[39] This also constitutes a natural boundary for many species of plants and animals and the true Antarctic life forms are found to the south of it.[40] An important international programme under the auspices of SCAR has been co-ordinating research in this region; it is called Biological Investigations of Marine Antarctic Systems and Stocks (BIOMASS) and is one of the largest international marine programmes ever mounted anywhere in the world.[41]

In these waters nutrients are not limiting, but levels of primary production are not significantly higher than in other oceans, possibly because the sustained strong winds make it the windiest and most turbulent of all oceans; because of this the primary producers – unicellular plants – are unable to maintain themselves at depths which enable them to make optimal use of energy from the sun. However the higher trophic levels – seabirds, seals and whales – are more abundant than in other oceans. This paradox may perhaps be explained by the nature of the as yet poorly understood pathways of nutrient cycling, seasonal limitation of trace elements and the key position of an unusually abundant and long-lived crustacean, the Antarctic krill.[42]

The Antarctic is also characterised by a strong seasonal pulse of primary production in summer and by the winter expansion of the pack-ice zone so as to cover more than half of the oceanic area south of the convergence.[43] The ecology of the pack-ice edge is important but still little studied, for logistic reasons; the pack-ice zone as a whole is even less well known. The extensive winter pack-ice cover may have a great influence on the behaviour of marine organisms, including the krill, with important implications for conservation and management of the resource. For these reasons and for its contribution to understanding marine ecosystems and processes elsewhere basic research in the Antarctic is important.

Many exciting research opportunities are being exploited concerned with the behaviour, ecology, physiology and biochemistry of krill.[44] These include large-scale studies of distribution, abundance and reproduction, swarming behaviour and its relation to feeding, experimental studies of swimming, feeding, energetics, swarming behaviour, moulting and activity patterns and biochemistry. Also research on krill is important because of the need for knowledge to manage the commercial fishery, which has not yet reached a significant size.[45]

Other components of the Antarctic marine ecosystem provide excellent opportunities for investigations that contribute to knowledge of marine ecosystems generally. Fish have received much attention, particularly in respect of their physiology and biochemistry, and adaptations such as the bloodless condition of certain species and the presence of antifreeze in their tissues and fluids.[46] The squids are another group of animals found in the Antarctic which are little known but on which research should be rewarding. Fish have been overexploited in the recent past[47] and squid may receive commercial attention as they have in recent years around the Falkland Islands, so improved knowledge is essential to their conservation and management.

Antarctic seabirds and seals have no fear of man and offer incomparable opportunities for studying behaviour, physiology and ecology.[48] Antarctic research continues to make very significant contributions in these fields; it is one of the longest running areas of Antarctic research and there are large populations of known-age individuals – banded, tagged or branded – giving excellent opportunities for longitudinal studies. Ecological studies of the role of seabirds and seals as predators on marine life, especially krill, are of particular importance. Examples of other topics that can probably be addressed better in the Antarctic than anywhere else are: population structure and dynamics, breeding success, social organisation and behaviour, feeding, energetics, and growth.

There are also unusual opportunities to study the processes involved in the dramatic recovery of the Antarctic fur seal from near extinction, the changes in elephant seal population ecology since commercial hunting ended, the increase of the crabeater seal, penguin and other seabird populations and of minke whale populations associated with the decline of commercial whaling.[49] These and other perturbations caused by the rise and decline of whaling and sealing offer great opportunities to study the results of such unplanned very large-scale experiments, which have a great deal to contribute to the general theory and understanding of ecosystem structure, function, and dynamics. Other discoveries in the Antarctic have contributed to research elsewhere, for example, the finding that annual layers in the teeth of seals enable us to determine their age[50] has been widely applied in research to other marine and terrestrial mammals.

Studies of the inshore marine ecosystems and their component organisms, while less urgent also have much to tell us; they also are subject to perturbations on a smaller scale from iceberg scouring.[11] They are more vulnerable to pollution and environmental impact from human activities, and we need to be able to predict the nature and importance of these impacts, so as to control and minimise them.

Life on land

Only a small fraction of the land area is not covered by ice and therefore capable of supporting higher forms of life; the area of inland waters is even more insignificant. But because they are relatively species-poor ecosystems, which have had minimal interference from man, studies of them facilitate the testing of hypotheses about biological strategies; also these are the areas most susceptible to environmental impact and so it is important to understand the processes involved.[12] There are no large herbivores, and relatively few species of microbes, plants and animals compared with other regions and a high level of endemism. Such systems provide ideal opportunities for investigating and modelling fundamental ecological processes – biogeochemical pathways, soil development, colonisation, succession, community development, and cyclical changes.[13]

The environment is much harsher and less stable than in the sea, with enormous seasonal fluctuation in air and soil temperatures. The aquatic environments of the inland waters constitute an exception, due to the insulating effect of ice and snow cover; some are permanently covered, others for perhaps eight months in the year, which converts them into more or less closed systems. They tend to be simpler than the terrestrial systems. Their great diversity resulting from the varying times available for development since the ice sheet receded, and their widely varying salinities, means that unusually rigorous comparative investigations are possible. The results here again are important for comparison with more complex systems elsewhere.[14]

Survival of individual organisms in these rigorous environments has produced a range of adaptive strategies affecting life histories, growth form, biochemical adaptation to desiccation and freezing, energetics, physiology, and behaviour. Again the principles involved can help to interpret other more complicated systems.

Last, but by no means least, personnel on Antarctic stations, tend to be within a narrow age band, physically fit, eating the same diet, and for much of the year they live in isolation within a confined area. It is almost impossible to find similar groups elsewhere in the world and this offers unique opportunities to follow some lines of medical and physiological research.

Conclusions

The uniqueness of the Antarctic environment has been widely recognised. Its fragile land and inland water ecosystems are very susceptible to disturbance and pollution, whereas the marine eco-systems are more stable and robust, even though they have been subject to large-scale perturbations from the sealing and whaling industries. There are no proven commercial mineral or hydrocarbon resources on land and none on the continental shelves; to assess them will require a very major effort, by geologists and geophysicists, at enormous cost, so the beginning of commercial activities is likely to be long delayed.

In both the short and long term the overall importance of the Antarctic to mankind is greater than purely commercial con-siderations. Its true value lies in the scientific opportunities provided by its unique position and environments, geography, geophysics, geology, glaciology, the ecosystems and their living organisms. Basic research on these unique conditions, features and organisms can give insights that can be applied to fundamental research elsewhere; SCAR has an important co-ordinating role. The values of the Antarctic fall into four main groups.

First, the insights, monitoring opportunities and indications of global change. Examples include the unexpected discovery of the ozone hole in spring over the Antarctic and decreasing levels of ozone world-wide; the contribution of the archive contained in the ice-cap relating to past climates and atmospheric constituents (gases, particles, heavy metals) which helps to understand and predict global warming; serial estimates of the changing volume of Antarctic ice, by integrating radio-echo sounding of bedrock contours and satellite altimetry of surface contours; research on mechanisms controlling the growth, flow and decay of the ice sheets in order to construct predictive mathematical models. The results are vital to under-standing and predicting a number of global changes which could have dire consequences for the human race. There is an important Antarctic component to the International Geosphere Biosphere Programme (IGBP) planned by ICSU.

Secondly, on a shorter timescale, the use of data collected in the Antarctic to contribute to global modelling and forecasting. Examples are: atmospheric modelling and weather forecasting; and the study of deep-space phenomena such as magnetic storms produced by gusts of the solar wind, by analysing their effects on the ionosphere and magnetosphere.

Thirdly, the study of patterns, processes and evolution, in geospace, in the atmosphere, in the lithosphere, the oceans and the biosphere. Examples are: studies of magnetic field lines; the ionosphere; radio wave generation and propagation; the breakup of Gondwana to form the present southern continents; plate tectonics, large-scale basic geology and solid-earth geophysics; palaeoclimates (palaeomagnetic, sediment core, ice core and palaeontological techniques); snow accumulation and ablation; ice flow and grounding lines; bottom-melting of ice shelves and formation of Antarctic Bottom Water; pack-ice expansion and contraction; current systems in the Southern Ocean; biogeochemical cycles, nutrient cycles and food webs on land, in lakes and at sea; biochemistry, physiology, behaviour and ecology of organisms, and the biological strategies involved in their survival.

Fourthly, applied research on actual or potential resources and environmental problems within the Antarctic. Examples are: assessment of krill, squid, fish, seal and whale population abundance; acquisition of knowledge necessary for the wise management of these resources; case studies of man-made perturbations and the introduction of alien flora and fauna; the environmental consequences of human activities (construction and waste disposal due to scientific research itself, mineral exploration and exploitation, pollution imported from industrial regions of the globe). Such research in the Antarctic has wider applications.

Up to the present, the Antarctic has been 'a continent for science' and this has served the world well politically, by minimising conflict in the region under the Antarctic Treaty system. Antarctic science has been very successful and a new role for it has emerged in relation to understanding and predicting global change, which is now becoming a major preoccupation of politicians. For these and other reasons plans for a different regime for the Antarctic must ensure that it allows for the continuing pursuit of scientific goals, with a minimum of restrictions. Unfortunately, some of the proposed regimes (for example, a World Wilderness Park, United Nations Trusteeship, the application of the ideas enshrined in the 'Common Heritage of

Mankind') are likely to militate against this, whether by weakening the Antarctic Treaty system, or in more specific ways.[55]

Notes

1 The following sources are particularly useful for the interested reader: W. N. Bonner and D. W. H. Walton (eds.), *Key Environments – Antarctica*, Pergamon Press, Oxford, 1985. C. Craddock (ed.), *Antarctic Geoscience*, University of Wisconsin Press, Madison, 1982. D. J. Drewry (ed.), *Antarctica: Glaciological and Geophysical Folio*, Scott Polar Research Institute, Cambridge, 1983. R. M. Laws (ed.), *Antarctic Ecology*, Academic Press, London, 1984. R. M. Laws, *Antarctica: The Last Frontier*, Boxtree, London, 1989. J. F. Lovering and J. R. V. Prescott, *Last of lands – Antarctica*, Melbourne University Press, Carlton, 1979. R. L. Oliver, P. R. James and J. B. Jago (eds.), *Antarctic Earth Science*, Australian Academy of Science, Canberra, and Cambridge University Press, 1983. D. W. H. Walton (ed.), *Antarctic Science*, Cambridge University Press, 1987.

2 W. Block, 'Terrestrial microbiology, invertebrates and ecosystems', in Laws, *Antarctic Ecology*, pp. 163–236; I. Everson, 'Antarctic food webs', in Walton, *Antarctic Science*, pp. 113–23; and D. W. H. Walton, 'Terrestrial environment', in Laws, *Antarctic Ecology*, pp. 1–60.

3 G. G. C. Claridge and I. B. Campbell, 'Physical geography – soils', in Bonner and Walton, *Key Environments – Antarctica*, pp. 62–70; and J. H. Zumberge (ed.), *Possible Environmental Effects of Mineral Exploration and Exploitation in Antactica*, SCAR, Cambridge, 1979.

4 R. M. Laws, 'Ecology of the Southern Ocean', *American Scientist*, LXXIII, 1985, pp. 26–40; and R. M. Laws, *Animal Conservation in the Antarctic*, Symposium LIV, Zoology Society, London, 1986, pp. 3–23.

5 Zumberge, *Possible Environmental effects of Mineral Exploration and Exploitation in Antarctica*.

6 D. H. Elliot, 'Antarctica: is there any oil and natural gas?', *Oceanus*, XXXI, 1988, pp. 32–7.

7 Sir Anthony Parsons, *Antarctica: The Next Decade*, Cambridge University Press, 1987.

8 Zumberge, *Possible Environmental Effects of Mineral Exploration and Exploitation in Antarctica*.

9 C. S. M. Doake, 'Glacial and climatic history', in Walton, *Antarctic Science*, pp. 151–60; and E. W. Wolff and D. A. Peel, 'The record of global pollution in polar snow and ice, *Nature*, CCCXIII, 1985, pp. 535–40.

10 A. L. Gordon, 'The Southern Ocean and global climate', *Oceanus*, XXXI, 1988, pp. 39–46.

11 J. R. Dudeney, 'Climate of extremes', in Walton, *Antarctic Science*, pp. 193–208.

12   R. Fifield, *International research in the Antarctic*, SCAR/ICSU Press, Oxford Science Publications, Oxford, 1987.

13   J. P. Kennett, 'Cenozoic evolution of Antarctic glaciation, the circum-Antarctic Ocean, and their impact on global palaeooceanography', *Journal of Geophysical Research*, LXXXII, 1978, pp. 3843–60.

14   Gordon, 'The Southern Ocean and global climate'; and T. Whitworth, 'The Antarctic circumpolar current', *Oceanus*, XXXI, 1988, pp. 53–8.

15   Doake, 'Glacial and climatic history'; and Dudeney, 'Climate of extremes'.

16   Gordon, 'The Southern Ocean and global climate'; and Whitworth 'The Antarctic circumpolar current'.

17   J. R. Dudeney, 'The Antarctic climate today' in Walton, *Antarctic Science*, pp. 209–21.

18   Dudeney, 'The Antarctic climate today'; and J. Gribbin, *The Hole in the Sky*, Corgi Books, London, 1988.

19   Gribbin, *The Hole in the Sky*; and J. Gribbin, *Hothouse Earth; the Greenhouse Effect and Gaia*, Bantam Press, London, 1990.

20   Dudeney, 'Climate of extremes'; and J. F. Farman, B. G. Gardiner and J. D. Shanklin, 'Large losses of total ozone in Antarctica reveal seasonal chlorine monoxide/nitrogen oxide interactions', *Nature*, London, CCCXV, 1985, pp. 207–10.

21   Gribbin, *The Hole in the Sky*.

22   Gribbin, *Hothouse Earth*.

23   G. de Q. Robin (ed.), *The Climatic Record in Polar Ice Sheets*, Cambridge University Press, 1983.

24   C. Lorius, *et al.* 'A 150,000 year climate record from Antarctic ice', *Nature*, CDXVI, 1985, pp. 591–6.

25   C. S. M. Doake, 'Sea ice and icebergs', in Walton, *Antarctic Science*, pp. 140–50.

26   Doake, 'Glacial and climatic history'.

27   J. R. Dudeney, 'Space research from Antarctica', in Walton, *Antarctic Science*, pp. 222–38.

28   Dudeney, 'Space research from Antarctica'.

29   *Ibid.*

30   D. J. Drewry, 'The Antarctic physical environment', in G. D. Triggs (ed.), *The Antarctic Treaty Regime: Law Environment and Resources*, Cambridge University Press, 1987, pp. 6–27; and D. H. Elliot, 'Physical geography – geological evolution', in Bonner and Walton, *Key Environments – Antarctica*, pp. 39–61.

31   Elliot, 'Physical geography–geological evolution; and Walton, 'The terrestrial environment'.

32   Craddock, *Antarctic Geoscience*; Elliot, 'Physical geography–geological evolution' and 'Antarctica: is there any oil and natural gas?'; and Drewry, 'The Antarctic physical environment'.

33 Drewry, 'The Antarctic physical environment'; Elliot, 'Physical geography–geological evolution' and 'Antarctica: is there any oil and natural gas?'.

34 Elliot, 'Antarctica: is there any oil and natural gas?'.

35 Craddock, *Antarctic Geoscience*; and I. W. D. Dalziel, 'The evolution of the Scotia Arc', in Oliver, James and Jago, *Antarctic Earth Science*, pp. 283–8.

36 Elliot, 'Physical geography–geological evolution' and 'Antarctica: is there any oil and natural gas?'.

37 Elliot, 'Physical geography–geological evolution'.

38 C. S. M. Doake, 'Keystone to Gondwana', in Walton, *Antarctic Science*, pp. 174–90; and Elliot, 'Physical geography–geological evolution'.

39 G. Deacon, *The Antarctic Circumpolar Ocean*, Cambridge University Press, 1984; and Whitworth, 'The Antarctic Circumpolar Current'.

40 Laws, 'Ecology of the Southern Ocean'; and S. Z. El-Sayed, 'Plankton of the Antarctic seas', in Bonner and Walton, *Key Environments – Antarctica*, pp. 135–53.

41 *Biological Investigations of Marine Antarctic Systems and Stocks (BIOMASS): Research Proposals*, BIOMASS Scientific Series, I, SCAR, Cambridge, 1977.

42 Everson, 'Antarctic food webs'; and El-Sayed, 'Plankton of the Antarctic seas'.

43 Doake 'Sea ice and icebergs'.

44 I. Everson, 'Physiological adaptation', in Walton, *Antarctic Science*, pp. 97–111; and D. Miller, *Biology and Ecology of the Antarctic Krill*, BIOMASS Scientific Series, IX, SCAR, Cambridge, 1989.

45 I. Everson, 'Exploitation of Antarctic fisheries', in Walton, *Antarctic Science*, pp. 125–37; and Miller, *Biology and Ecology of the Antarctic Krill*.

46 Everson, 'Physiological adaptation'; and K. H. Kock, 'Marine habitats – Antarctic fish', in Bonner and Walton, *Key Environments – Antarctica*, pp. 173–92.

47 K. H. Kock, G. Duhamel and J. C. Hureau, *Biology and Status of Exploited Antarctic Fish Stocks: A Review*, BIOMASS Scientific Series, VI, SCAR, Cambridge, 1985.

48 J. P. Croxall, 'Seabirds', and R. M. Laws, 'Seals', in Laws, *Antarctic Ecology*, pp. 533–620 and pp. 621–716 respectively.

49 Laws, 'Ecology of the Southern Ocean'; and J. Gulland, 'The end of whaling?' *New Scientist*, MDCXXXVI. 1988, pp. 42–7.

50 R. M. Laws, 'A new method of age determination for mammals', *Nature*, CLXIX, 1952, p. 972.

51 G. B. Picken, 'Marine habitats – benthos', in Bonner and Walton, *Key Environments – Antarctica*, pp. 154–72; and M. G. White, 'Marine benthos', in Laws, *Antarctic Ecology*, pp. 421–62.

52 Block, 'Terrestrial microbiology, invertebrates and ecosystems'; R. I.

Lewis Smith, 'Terrestrial plant biology', in Laws, *Antarctic Ecology*, pp. 61–162; R. E. Longton, 'Terrestrial habitats – vegetation; and L. Somme, 'Terrestrial habitats – invertebrates', in Bonner and Walton, *Key Environments in Antarctica*, pp. 73–105 and pp. 106–17 respectively.

53 Block, 'Terrestrial microbiology, invertebrates and ecosystems'; R. I. Lewis Smith, 'Terrestrial plant biology', in Laws, *Antarctic Ecology*, pp. 61–162.

54 R. B. Heywood, 'Inland Waters', in Laws, *Antarctic Ecology*, pp. 279–344; and J. Priddle, 'Terrestrial habitats – inland waters', in Bonner and Walton, *Key Environments – Antarctica*, pp. 118–32.

55 R. M. Laws, 'Science, the Treaty and the future', in Walton, *Antarctic Science*, pp. 250–65.

The mineral resource potential of Antarctica: geological realities

Introduction

Antarctica, with an area of fourteen million square kilometres, is the fifth largest continent on earth. Most of it is covered by ice sheets with a mean thickness of 2,200 metres (and a maximum of 4,800 metres), through which protrude numerous mountain ranges up to 5,100 metres in height. Although overall rock exposure is less than two and a half per cent, it is far greater in the mountains and some coastal regions. Geologically, the continent is the least explored in the world. Of the 332,000 square kilometres of rock not covered by ice, about eighty per cent has been geologically mapped at a reconnaissance level (about 1:500,000 scale), possibly ten per cent in detail (1:50,000 scale or larger). Many large areas have not been revisited since they were first surveyed.

The Convention on the Regulation of Antarctic Mineral Resource Activities (CRAMRA) has been negotiated, therefore, in the absence of knowledge of what types of mineral concentration might exist in Antarctica. The inaccurate use of terms such as 'mineralisation', 'mineral' (or 'ore') 'deposits' and 'reserves' has given rise to the belief that successful mineral exploration has occurred and that economic deposits have been defined. The popular media have further exaggerated these reports and, with non-governmental organisations, have suggested that CRAMRA will clear the way for mining operations to start on the Antarctic continent.

It is crucial in this context to define what is meant by 'Antarctic mineral resources' and to assess, as far as possible, the geological evidence for the mineral potential of the continent and its surrounding shelf. For 'hard minerals' (excluding the hydrocarbons), emphasis is placed on the Antarctic Peninsula; it is one of the best exposed and most studied regions, and is thought by many to be the most mineralised

For hydrocarbons, current knowledge of the whole continent, both continental margins and supposed interior basins, is assessed.

It should be understood that the philosophical basis of exploration for hydrocarbons and minerals is different. Hydrocarbon reservoirs occur at depth in the crust, therefore, their discovery and definition is always an expensive and lengthy process. In contrast, hard mineral occurrences on land may first be recognised. However, compared to proven hydrocarbon reservoirs, the size and quality of hard mineral occurrences are more unpredictable. Hence their exploration, definition and beneficiation are also costly and time consuming.

Geological research in Antarctica

Systematic regional geological mapping of Antarctica began only after 1945. In the last fifteen years most effort has been directed towards problem-oriented research on fundamental geological processes.[1] During mapping (generally by geologists without specialist economic geology experience) 'mineralised' rocks have been reported at several hundred isolated localities. Few of these have been revisited and rarely have they been mapped, sampled and analysed in the detail required to demonstrate either that trace element **anomalies** exist or that mineralising **processes** have occurred. Despite the lack of data, geological papers have quoted 'deposits',[2] 'economic deposits'[3] and 'ore deposits'.[4] Many reviews and the media further mislead by stating that certain areas are **known** to be rich in mineral resources.[5] Clearly, the precisely defined terms of economic geology must be explained.

What are mineral resources?

Although the earth's crust is composed dominantly of a small group of elements, contained in a limited number of mineral and rock types, trace element levels are extremely variable.[6] Some rock types may contain major and trace elements, enriched ('mineralised') or depleted, from normal crustal levels by a variety of common magmatic, sedimentary or metamorphic processes, or rare hydrothermal–metasomatic processes.

A **mineral occurrence** is a locality where the rocks contain an anomalous mineralogy and geochemistry. Such occurrences may be a few millimetres to several kilometres across. They are ubiquitous in all parts of the continental and oceanic crust, including Antarctica.

Mineral occurrences are found by accident or by expensive exploration programmes. The term 'mineral occurrence' may also be used for a locality (or **prospect**) which has been explored in detail, but whose location, geological structure, mineralogy, geochemistry or size are inadequate to make it a mineral **deposit** and hence does not allow profitable exploitation in the foreseeable future.

Mineral deposits are volumes of rock whose geology, mineralogy, geochemistry and size allow extraction of a valuable commodity at a profit. The term may include rocks useful for their content of metals, non-metals, oil or gas, and rocks used in their entirety (coal and the bulk minerals). Hard mineral deposits are defined by detailed programmes of geological and geophysical mapping, excavation, drilling and laboratory analysis, followed by exploratory mining, bulk sampling, mineral beneficiation and economic studies. Discovery statistics from North America, South Africa and Australia suggest that of one hundred mineral occurrences, three might be prospects examined in detail and one may be a significant mineral deposit. The appraisal process is similar for hydrocarbons, except that the very high cost of offshore operations means that geophysical exploration is stressed before drilling. On a world-wide average, about one wildcat well in ten discovers hydrocarbons in economic quantities. Successful exploration programmes for both hard minerals and hydrocarbons may last five to ten years, or longer, with a very high outlay for each significant discovery. There are no mineral deposits known in Antarctica because no such definition work has been carried out.

Economic deposits are volumes of rock being extracted, processed and sold at a profit under favourable logistic, economic and political conditions. These vary tremendously from one political/market system to another. In hard rock mines, the material conveyed to the mill is known as **ore. Reserves** are volumes of rock identified as the ore or oilfields of tomorrow.

Mineral resources are a summation of hypothetical deposits and mineral reserves which have been identified. The whole of the earth's crust, including Antarctica, contains mineral resources (*see* figure 3.1).

Mineral resource potential is the measure of the possibility of rocks occurring which are of potential use to man. A terrane[7] with a high potential is one where geological, geochemical and geophysical indications are anomalous, where occurrences are known and deposits have been defined. Moderate potential is where indications are anomalous, occurrences are known but no deposits have been defined.

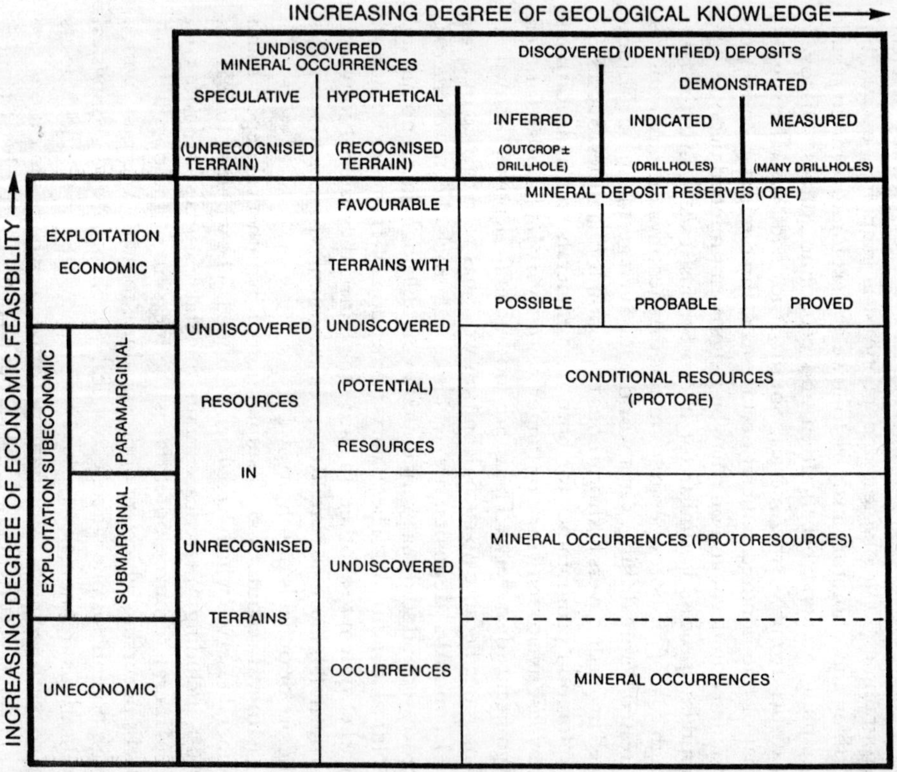

Figure 3.1 Definition of mineral reserves and resources in terms of geological knowledge and economic feasibility (the latter related to grade, tonnage and numerous logistic, technological, economic and political considerations)

Low potential is where there are no anomalous indications or known occurrences. An unknown potential is where the data available are inadequate to make an assessment. No area of the earth's crust has a potential of zero.

Mineral resource assessments are geological syntheses to determine the mineral resource potential.

For hard minerals in the inhabited continents, the database assessed would include regional geological maps, at scales better than 1:250,000 (often at 1:50,000), regional stream sediment geochemistry typically at one sample per square kilometre, and airborne geophysical data collected at 100 metres ground clearance and 500 metres line spacing. Stream sediment sampling is a way of screening large areas for geochemical anomalies, but is not possible in the Antarctic due to its heavily glaciated, mountainous terrain. The thick ice sheets also mask magnetic or electrical anomalies induced by mineralised rocks, hampering geophysical surveys. In addition, the ground clearance and line spacings of the surveys which have been carried out (typically 2,500 metres and 20 kilometres respectively)[8] are far greater than those required to detect anomalies due to mineralised rocks. To summarise, the only method of discovering, mapping and studying mineralised rocks in Antarctica is by direct, detailed examination of every exposure. This relatively expensive method of exploration is further handicapped by the absence of the conspicuous indicators of mineralisation used as exploration tools elsewhere in the world, notably secondary enrichments and gossans.

For offshore hydrocarbons the major tool, both in exploration and assessment, is the marine seismic survey; only after the most thorough survey is any drilling undertaken. In mature hydrocarbon provinces such surveys are carried out on grids with a spacing of 500 metres; the most detailed seismic survey carried out in Antarctic waters to date (for strictly scientific reasons) had a line spacing of more than 100 kilometres.[9]

For both hard minerals and hydrocarbons, assessments are also based on the theoretical, empirical, genetic and exploration models derived from elsewhere. Due to changing fashions in modelling, final authoritative assessments of any part of the earth's crust cannot be made. Assessments cannot predict the exact location of undiscovered mineralisation, nor do they take into account exploration costs or the geotechnical, economic or political factors which may change the definition of what is economic.

Mineral resource assessment of Antarctica

An assessment of Antarctica may be attempted broadly along the lines
indicated above, by first comparing the existing geological database
with exploration models derived elsewhere; secondly, by compiling
and assessing the known mineral occurrences and geochemical
anomalies; and thirdly, by comparison with once contiguous terranes
which are more accessible and hence better known.

The geology of Antarctica

East Antarctica is similar in size to Australia; it consists of a
metamorphic craton, dated between 3,800 and 800 million years old
(*see* figure 3.2). The Transantarctic Mountains form a 3,000 kilometre-
long belt of deformed continental margin sedimentary and igneous
rocks, of late Precambrian to early Palaeozoic age (750 to 500 million
years old), overlain by relatively undeformed continental sedimentary
and igneous rocks of Palaeozoic to Mesozoic age (400 to 180 million
years old).

West Antarctica consists of a mosaic of five or more separate crustal
blocks. Recent fieldwork has shown that each block has distinct
geological elements,[10] although their extent is the subject of current
research. Palaeomagnetic results suggest that the blocks are all of local
derivation.[11] The Antarctic Peninsula lies on one of the two largest
blocks. It consists of Palaeozoic basement and Palaeozoic–Mesozoic
subduction complex, overlain by volcanic and plutonic rocks formed
in a Mesozoic–Cenozoic island arc (180 to 10 million years old).
Marine sedimentary basins formed on the flanks of the Antarctic
Peninsula arc, and there are widespread rift-related volcanic rocks of
Cenozoic age.

The geological database

The geological terranes identified may host a wide variety of
geochemically anomalous rocks, formed either by normal magmatic,
sedimentary or metamorphic processes[12] or by fluid movement in the
crust.[13] However, the available geological maps are not sufficiently
detailed to identify promising settings for mineralisation such as
volatile-rich intrusions, central stratovolcano and caldera complexes,
submarine volcanic sequences or crustal lineament intersections.

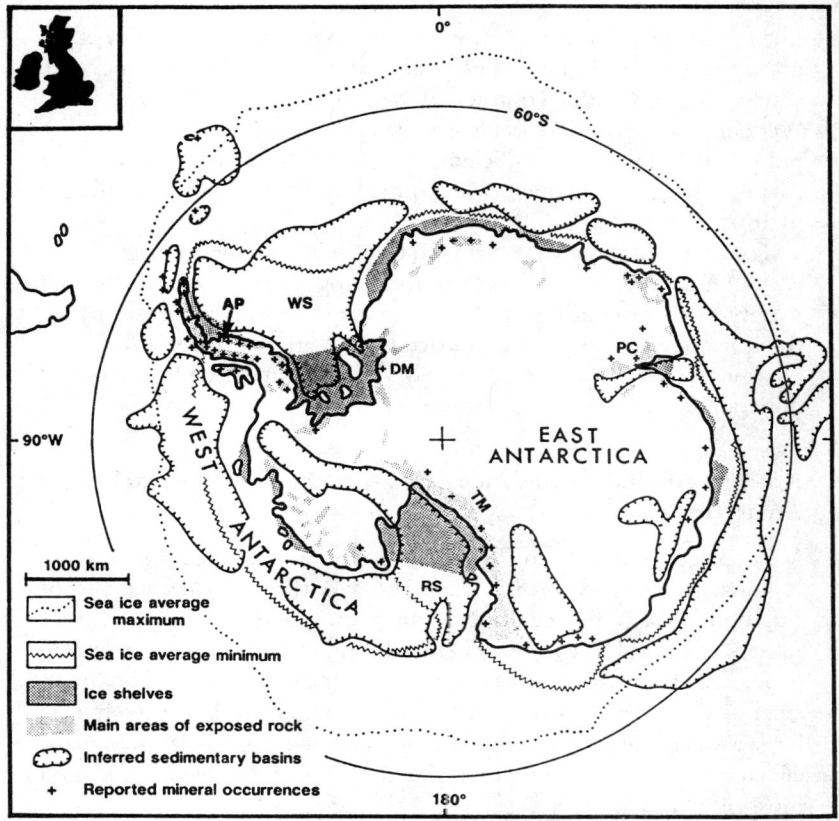

Figure 3.2 A map of Antarctica showing the extent of the ice shelves, seasonal pack-ice, distribution of exposed rock, reported on-land mineral occurrences (generalised) and postulated sedimentary basins. The key to the abbreviations is: AP – Antarctic Peninsula; DM – Dufek Massif; PC – Prince Charles Mountains; RS – Ross Sea; TM – Transantarctic Mountains; WS – Weddell Sea.

Offshore sedimentary basins are from St John[33] and onshore sedimentary basins modified from Drewry.[34] Inset map shows the British Isles at the same scale.

For hydrocarbons, existing geophysical surveys have done little more than delineate the position of the major Mesozoic–Cenozoic sedimentary basins (*see* figure 3.2). Some basins are inferred merely from present-day topography and passive geophysical methods such as gravity and magnetic surveys. There have been no significant seismic surveys of the internal structure of the on-land basins to date, while marine seismic reflection lines (and a very few refraction lines) are scattered around the continental margin. Only in the Ross and Weddell seas have there been any 'detailed' scientific seismic studies published.[14] These are sufficient to identify the major sedimentary basins, but are inadequate for defining structures which could be hydrocarbon traps.

To summarise, there is no reason to believe that the Antarctic crust is not mineralised or hydrocarbon-bearing to the same extent and variety as other continents. However, these resources remain hypothetical and for the most part buried beneath the ice sheet. Like other continents, the Antarctic crust may also contain unknown types of mineralisation (speculative resources).

Known mineral occurrences and geochemical anomalies in Antarctica

Mineralisation has been reported at about 300 localities during regional mapping over forty years.[15] Most of these localities consist of small outcrops (a few centimetres to metres) and for many localities there is little evidence for geochemical anomalies. However, in review papers and the media, their geographic distribution is often shown on small-scale maps as symbols whose size suggest authoritatively that large areas of mineralisation have been defined. In reality, such maps effectively define the areas of exposed rock which have been visited most often (*see* figure 3.2).

The most widely quoted rocks are the banded iron formations (BIF) of the Prince Charles Mountains. Although extensive, they are not large compared to BIF elsewhere, they are also relatively low grade (thirty-five per cent iron) and contain high contaminant phosphorus pentoxide. Similarly, the widespread Permian coal-bearing formations of the Transantarctic and Prince Charles Mountains are commonly cited as a potential resource.[16] However, other geologists point out that field reports of coal seam thicknesses are almost certainly exaggerated as they include shale bands.[17] They conclude that data are too scattered to make even the broadest resource assessment of Antarctic coal and doubt any possibility of economic coal exploitation. Being located hundreds of kilometres inland from ice coasts which are accessible for only two months each year, it is unlikely that Antarctic iron or coal could compete with the thousands of millions of tonnes of more accessible and higher-grade iron and coal resources which remain unexploited in other parts of the world.

Approximately twenty other types of metallic mineral, and fifteen non-metallic minerals and rocks, used elsewhere as 'bulk minerals', have been reported in Antarctica. A few localities have been studied from the academic viewpoint during regional mapping but there has been no detailed definition work. The region with the highest concentration of localities, the Antarctic Peninsula, has 250 reported localities over an area of 400,000 square kilometres (compared to the British Isles with more than 4,000 localities in 241,000 square kilometres). Although the region studied in most detail, only nine papers on individual localities have been published in the international literature since 1975.[18] The styles of mineralisation described include mafic intrusion-hosted stratiform iron and chromium; porphyry-hosted copper and molybdenum; pegmatite copper, molybdenum and manganese; volcanogenic iron and copper; and epithermal-hypo-thermal vein and breccia copper, zinc, lead and molybdenum. Anomalous concentrations of various metals have been measured in samples from twenty-six localities out of eighty analysed. However, their significance is uncertain as background levels for the Antarctic Peninsula have not been established.

It has been suggested that the reported mineralisation in the Antarctic Peninsula occurs in north–south clusters related to lateral segmentation of the magmatic arc, and that some clusters show an east-west compositional zoning comparable to the heavily mineralised central Andes.[19] A detailed compilation of all unpublished and published data for the Antarctic Peninsula region suggests that the reported localities are clustered in areas of relatively easy access which have had a longer history of geological investigation. Large areas apparently lacking mineralisation have, in fact, not been visited or mapped. This suggests that sampling bias is the cause of the lateral clustering.

Detailed field examination of about forty localities in the Antarctic Peninsula region suggests that two-thirds of reported iron and copper localities are probably innocuous, consisting of accessory magmatic minerals, or (together with manganese) are visually conspicuous secondary mineral patinas formed by weathering of ferromagnesian silicates and accessory opaques in the cold arid climate.[20] Hand specimens from many localities, some described as 'porphyry copper deposits' contain from twenty (the crustal average) to a few hundred parts per million of copper, similar to many unmineralised basic and intermediate igneous rocks. Although there has been much talk of 'porphyry-style copper and molybdenum mineralisation', the few localities confirmed consist of joint-, fracture- and vein-hosted iron, copper or molybdenum sulphides in medium-grained equigranular granodiorites. The textural features and alteration patterns characteristic of porphyry-style mineralisation have not yet been observed or described in the literature. Other localities are mineral misidentifications, commonly galena or sphalerite for hematite. In contrast to the above negative results, detailed mapping in at least three, previously well-described, areas has revealed examples of previously unrecognised mineralisation.

In the area of reported east-west Andean-type 'zoning', only six out of twenty-eight localities examined were confirmed. This may be due to the vagaries of fieldwork and the fallibility of human perception. However, the lack of true porphyry mineralisation, and discovery of new localities in the same area, do not instil confidence in lateral zoning models and hence make comparisons with the central Andes premature.

A cursory literature study for the rest of Antarctica suggests that reported titanium, uranium, thorium and tin 'mineral occurrences' and vanadium, zirconium, chromium, nickel, cerium, cobalt, lanthanum and niobium 'anomalies' are of uncertain significance. These minerals and elements may be naturally ubiquitous or occur at high levels in some rock types. The 'platinum deposits' (along with iron, vanadium, chromium, nickel and cobalt) of the Dufek layered basic intrusion[21] are hypothetical.[22] No platinum-bearing minerals or whole-rock platinum anomalies have been reported in these rocks.

In conclusion, the Antarctic continent has a small number of mineral occurrences and areas of geochemical anomaly. Their apparent distribution is biased by topographic and exploration reasons. For many other localities, there is little evidence for either whole rock anomalies or mineralisation processes.

Antarctica and continental reconstructions

Since the advent of the theories of continental drift and plate tectonics, comparisons of different regions on continental reconstructions has been a popular method of 'resource suggestion'.

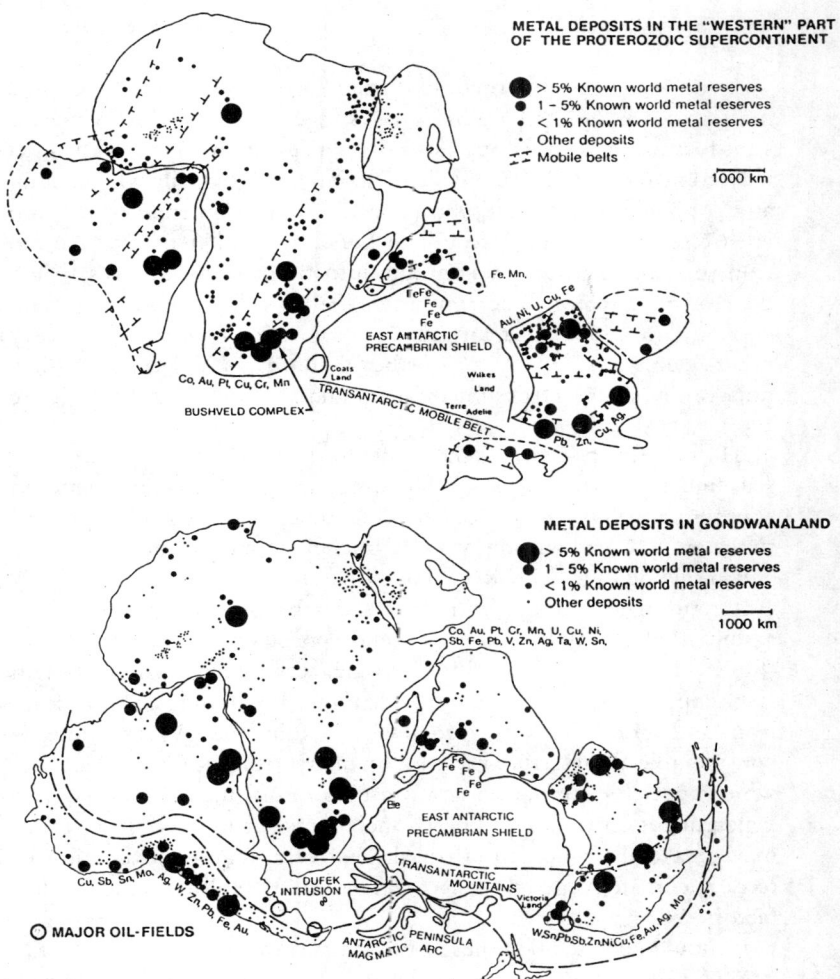

METAL DEPOSITS IN THE "WESTERN" PART
OF THE PROTEROZOIC SUPERCONTINENT

● > 5% Known world metal reserves
● 1 – 5% Known world metal reserves
• < 1% Known world metal reserves
· Other deposits
⊐⊏ Mobile belts

1000 km

Fe, Mn.

FeFe
Fe
Fe

Au, Ni, U, Cu, Fe

EAST ANTARCTIC
PRECAMBRIAN SHIELD

Coats
Land

Wilkes
Land

Co, Au, Pt, Cu, Cr, Mn

TRANSANTARCTIC MOBILE BELT

Terre Adelie

BUSHVELD COMPLEX

Pb, Zn.

Cu, Ag.

METAL DEPOSITS IN GONDWANALAND

● > 5% Known world metal reserves
● 1 – 5% Known world metal reserves
• < 1% Known world metal reserves
· Other deposits

1000 km

Co, Au, Pt, Cr, Mn, U, Cu, Ni,
Sb, Fe, Pb, V, Zn, Ag, Ta, W, Sn.

Fe
Fe
Fe

Be

EAST ANTARCTIC
PRECAMBRIAN SHIELD

TRANSANTARCTIC
MOUNTAINS

DUFEK
INTRUSION

Cu, Sb, Sn, Mo.

Ag, W, Zn, Pb, Fe, Au.

Victoria
Land

○ MAJOR OIL-FIELDS

ANTARCTIC PENINSULA
MAGMATIC ARC

W,Sn,Pb,Sb,Zn,Ni,Cu,Fe,Au,Ag, Mo

Figure 3.3 Continental reconstructions for (a) the period 2,500 to 600 million years ago showing the position of the Antarctic craton in relation to Africa, Madagascar, India, Australia and north China[35] and (b) the period up to 180 million years ago showing the former relationship of Antarctica to the other southern continents.[36] Publicised metal deposits are indicated by solid symbols and major oil fields (in the immediate vicinity of Antarctica only) by shaded symbols. The key to the chemical symbols is as follows: Ag – silver; Au – gold; Be – beryllium; Co – cobalt; Cr – chromium; Cu – copper; Fe – iron; Mn – manganese; Mo – molybdenum; Ni – nickel; Pb – lead; Pt – platinum; Sb – antimony; Sn – tin; Ta – tantalum; Th – thorium; U – uranium; V – vanadium; W – tungsten; Zn – zinc.

A continental reconstruction, for the Proterozoic Era (2,500 to 590 million years ago; *see* figure 3.3a) suggests that the province of stratiform zinc, lead, copper and silver deposits of central Australia may continue into Terre Adélie. The Yilgarn gold, nickel, uranium and copper province (and about twenty other elements) of western Australia may continue into Wilkes Land. The Indian manganese and iron province appears to continue into the seventy degrees to ten degrees East sector. The southern African cobalt, gold, platinum, copper, chromium and manganese province (also with about twenty other elements), if on a 'northeast–southwest' lineament, may not continue into Dronning Maud Land or Coats Land (*see* figure 3.3a).

The Gondwana refit, for the early Mesozoic (180 million years ago; see figure 3.3b), places the Transantarctic Mountains between southern South Africa and eastern Australia, the latter with important tungsten, tin, lead, antimony, zinc, nickel, copper, iron, gold, silver and molybdenum deposits.[23] The apparent continuity of the West Antarctic magmatic arc with the Andes has led many workers to assume that it may host similar metal deposits to the central Andes (fifteen degrees South to thirty-five degrees South), which contain large copper, antimony, tin, molybdenum, silver, tungsten, zinc, lead, iron and gold deposits. However, within the Andes there is considerable along-strike variation in many geological features, all of which affect the formation and preservation of metal deposits. The apparent concentration of known metal deposits in the central Andes may be a real phenomenon or, alternatively, may be a function of terrain, climate, vegetation, access, amount of exploration and other factors.[24]

It should be noted that these continental comparisons are simplistic. The continents not shown in figure 3.3, namely, North America, Europe and Asia, are also mineral-rich and contain two-thirds of the world's known metal reserves. Whatever refit is used, Antarctica will appear to be surrounded by mineral-rich continents. The Gondwana comparison states the obvious – that the Antarctic crust is probably as mineralised as the other continents.

Antarctic hydrocarbons

No oil or gas seeps are known from the Antarctic, nor has any potential hydrocarbon-bearing structure been identified, either by geophysical surveying or geological mapping. There are only two hydrocarbon shows reported from the Antarctic. The first was of gas from Miocene and Pliocene strata at Deep Sea Drilling Project Sites 271, 272 and 273 in the Ross Sea. These samples may be thermogenic, but there is also considerable biogenic methane.[25] The second occurrence was from an eight-metre core from surface sediments of the Bransfield Strait (north of the Antarctic Peninsula); this contained methane–pentane in concentrations up to 350 parts per billion.[26] These were thought to be thermogenic hydrocarbons caused by high local heat flow, associated with contemporaneous volcanism.

In the absence of data, all published 'resource assessments' for the Antarctic are hypothetical. The main appraisals, and comprehensive reviews of previous literature are given by Behrendt[27] and St John,[28] who both considered the hydrocarbon potential of the whole continent. Behrendt plotted all known giant hydrocarbon fields on a Gondwana reconstruction at 180 million years ago (*see* figure 3.3b) and suggested that West Antarctica and the Ross and Weddell seas might be more prospective than the shelf of East Antarctica. He came to no firm conclusions regarding the hydrocarbon resource potential. St John reviewed all available literature to provide a map of the sedimentary basins of the Antarctic (*see* figure 3.2). He calculated the volume of each of the twenty-one basins identified and applied a world-wide average of 12,000 barrels of oil or oil-equivalent (BOE) per cubic kilometre of sedimentary rock to give a potential hydrocarbon yield of 203 billion BOE for the whole continent.

Despite their different methods, both authors argued by analogy, applying exploration data and models derived elsewhere to the Antarctic situation. This approach is seriously flawed, as it ignores geological detail. Even small changes in the geology between two basins can alter radically their relative prospectivities.[29] In particular, the Gondwana analogy is untenable, as the giant oil-fields in southern South America and southern Australia were formed *after* the breakup of the supercontinent, and reflect purely local geological events.

The most detailed assessment to date considered all the geological and geophysical evidence relating to the Larsen Basin, northeast of the Antarctic Peninsula.[30] The authors concluded that this large basin had moderate potential for oil generated from Upper Jurassic source rocks, reservoired in Cretaceous and Tertiary sandstones. However, the area of interest is among the most difficult to access in the Antarctic.[31]

Factors influencing possible development of Antarctic minerals

The above review suggests that Antarctica has an unquantifiable mineral resource potential. However, on land, significant mineral occurrences may be discovered by accident or design at any time. Whether these are explored or developed into reserves will depend upon a range of complex economic, political and environmental factors.

The resource base of the developed world

It has always been assumed that depletion of known reserves leads to an increase in demand and price; hence, at some future time, Antarctica will be exploited. However, the assumption that the inhabited continents can be depleted of resources may be erroneous. First, the total volume of crust already exploited is minuscule compared to the volume of unexplored crust in accessible regions. Second, availability of resources increases with improvements in predictive exploration models and technology. Third, the resource base can be changed by conservation, recycling and the development of substitute materials. If commodity prices do rise, mineral occurrences and economically marginal deposits in the inhabited continents are likely to have lower exploration and exploitation costs than the less well known rocks of Antarctica.

The Arctic analogue

Humans have inhabited the Arctic for thousands of years. Prospecting and exploration in this frontier region have led to the discovery of major reserves of hydrocarbons and minerals. Recent estimates place the level of potentially recoverable hydrocarbons at 100 to 200 billion barrels of oil and 50 to 100 trillion cubic metres of gas. Alaska possesses four trillion tonnes of coal reserves (equivalent to all other reserves of coal elsewhere in the USA) while estimates of Soviet northern coal reserves are of the order of seven trillion tonnes. Numerous large metal deposits (defined as containing more than fifty million tonnes) have been delineated. For example, the Red Dog deposit in northwest Alaska, proved by a ten-year exploration programme including eleven kilometres of drilling, contains seventy-seven million tonnes of twenty-two per cent zinc and lead, and eighty-two grammes per tonne of silver. These proven reserves stand in stark contrast to our knowledge of Antarctica.

Nevertheless, the hostile natural conditions, remoteness, and lack of infrastructure make exploration for and exploitation of Arctic resources very costly. Exploration wells in the Beaufort Sea can cost fifty times as much as those in the Gulf of Mexico. Many large deposits in the Arctic, which would be exploited elsewhere, are economically marginal. For instance, the giant Amauligak oilfield in the Beaufort Sea, with proven reserves of one billion barrels, represents the minimum size to justify development. Exploration and exploitation costs in Antarctica are likely to be considerably higher than in the Arctic. Antarctica has no settled population, no roads, one or two rudimentary ports and airstrips, a primitive communications system, no readily available labour pool, no sources of power, no cultural amenities and no local markets.

Geopolitical factors

It has been argued that resource exploitation in the Arctic is driven by its economic and political attractiveness compared to other politically less stable regions of the world.[32] The USA, Canada and USSR can explore for and develop resources in huge areas which are secure and politically predictable. Furthermore, circum-Arctic governments, besides exercising exclusive management authority over resources, may offer potent financial incentives, which are not based solely upon market forces, to encourage (or discourage) development.

In Antarctica, by contrast, demarcation of sovereignty and mineral rights have been left open under the terms of the Antarctic Treaty. There is added uncertainty as to how the Treaty regime will evolve during the 1990s, particularly given the current debate over CRAMRA. Although the CRAMRA provisions do not specify economically viable or logistically feasible exploitation, they require strict compliance with a number of measures. These include a ban on activity in environmentally sensitive zones (for example, Specially Protected Areas, Sites of Special Scientific Interest); consensus to agree the opening of an area for exploration or development; stringent procedures to protect the environment during exploration and/or development; proof of technically feasible and safe procedures; demonstration of compatibility with other legitimate uses (especially scientific) of Antarctica; monitoring and random inspection of activities; and liability and penalty provisions for environmental damage.

However should CRAMRA not be ratified, uncertainty will increase, as there will be no legal basis for mineral activity. This may lead to renewed political tension over sovereignty, particularly in the sector of UK–Chile–Argentina territorial counter-claims twenty degrees West to ninety degrees West. Antarctica will thus join the ranks of those parts of the world designated 'unstable' for mineral development because of 'internal' political uncertainty.

These circumstances would be in addition to the geological evidence reviewed here which shows that at present Antarctica contains no more than hypothetical and speculative resources.

Acknowledgements

The authors are grateful to Drs T. Alabaster, B. C. Storey and M. R. A. Thomson for reading the manuscript, and to Mr A. C. Sylvester for drafting the diagrams.

Notes

1   See, for example, Natural Environment Research Council, *Antarctica 2000: NERC strategy for Antarctic Research*, Natural Environment Research Council, Swindon, 1989.

2   For example, P. D. Rowley and W. R. Vennum (eds.), *Studies of the Geology and Mineral Resources of the Southern Antarctic Peninsula and Eastern Ellsworth Land, Antarctica, US Geological Survey Professional Paper, 1351*, Reston, Virginia, 1988.

3   For example, M. Pincheira, M. Pereira, G. Hoecker and E. Abad, 'Geological antecedents and mineralization occurrences in the southern sector of Gerlache Strait, Antarctic Peninsula', *Serie Científica INACH*, XXXIV, 1986, pp. 9–22.

4   For example, A. Paulo and Z. Rubinowski, 'Pyrite mineralization on King George Island (South Shetland Islands, Antarctica): its distribution and origin', *Studia Geologica Polonica*, XL, 1987, pp. 39–86.

5   See, for example, K. S. Jayaraman, 'New Indian base planned on Weddell Sea', *Nature*, CCCXLII, 1989, p. 845.

6   K. H. Wedepohl (ed.), *Handbook of Geochemistry*, Springer-Verlag, Berlin, vol. 1, 1978.

7   A discrete block of the earth's crust with a geological history distinct from neighbouring blocks.

8   For example, R. G. B. Renner, L. J. S. Sturgeon and S. W. Garrett, *Reconnaissance gravity and aeromagnetic surveys of the Antarctic Peninsula, British Antarctic Survey Scientific Report, No. 110*, Cambridge, 1985.

9 A. K. Cooper and F. J. Davey (eds.), *The Antarctic continental margin: geology and geophysics of the western Ross Sea*, Circum-Pacific Council for Energy and Mineral Resources, Houston, Texas, 1987.

10 B. C. Storey, I. W. D. Dalziel, S. W. Garrett, A. M. Grunow, R. J. Pankhurst and W. R. Vennum, 'West Antarctica in Gondwanaland: crustal blocks, reconstruction and breakup processes', *Tectonophysics*, CLV, 1988, pp. 381–90.

11 A. M. Grunow, D. V. Kent and I. W. D. Dalziel, 'Evolution of the Weddell Sea basin, new palaeomagnetic constraints', *Earth and Planetary Science Letters*, LXXXVI, 1987, pp. 16–26.

12 See D. P. Cox and D. A. Singer, 'Mineral deposit models', *US Geological Survey Bulletin, 1693*, Washington, DC, 1986.

13 See, for example, the following papers: J. M. Ferry, 'Contact metamorphism of roof pendants at Hope Valley, Alpine County, California, USA', *Contributions to Mineralogy and Petrology*, CI, 1989, pp. 402–17. B. S. Sibbet, 'Size, depth and related structures of intrusions under strato-volcanoes and associated geothermal systems', *Earth Science Reviews*, XXV, 1988, pp. 291–309; and P. Vrolijk, 'Tectonic driven fluid flow in the Kodiak accretionary complex, Alaska', *Geology*, XV, 1987, pp. 466–9.

14 See K. Hinz and M. Block, 'Results of geophysical investigations in the Weddell Sea and in the Ross Sea, Antarctica', *Proceedings of the 11th World Petroleum Congress 2*, John Wiley and Sons, Chichester, 1984, pp. 79–91' and Cooper and Davey, *The Antarctic Continental Margin*.

15 See R. C. R. Willan, 'Metalliferous mineralization in the Antarctic Peninsula', *Proceedings of the Pacific Rim Congress*, 87, The Australasian Institute of Mining and Metallurgy, Victoria, 1987, pp. 483–6; and P. D. Rowley, P. L. Williams and D. E. Pride, 'Minerals occurrences of Antarctica', in J. C. Behrendt (ed.), *Petroleum and mineral resources of Antarctica, US Geological Survey Circular, 909*, Reston, Virginia, 1983, pp. 25–49.

16 N. A. Wright and P. L. Williams, *Mineral Resources of Antarctica, US Geological Survey Circular, 705*, Reston, Virginia, 1974.

17 G. Rose and C. T. McElroy, 'Coal potential of Antarctica', *Bureau of Mineral Resources: Resource Report 2*, Canberra, 1987.

18 These papers are: P. D. Rowley, P. L. Williams, D. L. Schmidt, R. L. Reynolds, A. B. Ford, A. H. Clark, E. Farrar and S. L. McBride, 'Copper mineralization along the Lassiter Coast of the Antarctic Peninsula', *Economic Geology*, LXX, 1975, pp. 982–92; P. D. Rowley, P. L. Williams and D. L. Schmidt, 'Geology of an Upper Cretaceous copper deposit in the Andean Province, Antarctic Peninsula', *US Geological Survey Professional Paper, 984*, 1977; P. D. Rowley and D. E. Pride, 'Metallic mineral resources of the Antarctic Peninsula', in C. Craddock (ed.), *Antarctic Geoscience*, University of Wisconsin Press, Madison, 1982, pp. 859–70; Rowley, Williams and Pride, 'Minerals occurrences of Antarctica'; Rowley and Vennum (eds.), *Studies of the Geology and Mineral Resources of the Southern Antarctic Peninsula*; P. D.

Rowley, E. Farrar, P. E. Carrara, W. R. Vennum and K. S. Kellogg,
'Porphyry-type copper deposits and potassium–argon ages of plutonic rocks
of the Orville Coast and eastern Ellsworth Land, Antarctica', in Rowley and
Vennum (eds.), *Studies of the Geology and Mineral Resources of the Southern
Antarctic Peninsula and Eastern Ellsworth Land Antartica*, pp. 35–9; D. D.
Hawkes and M. J. Littlefair, 'An occurrence of molybdenum, copper and
iron mineralization in the Argentine Islands, West Antarctica', *Economic
Geology*, LXXVI, 1981, pp. 898–904; M. J. Littlefair, 'The "quartz–pyrite"
rocks of the south Shetland Islands, western Antarctic Peninsula', *Economic
Geology*, LXXIII, 1978, pp. 1184–9; and C. Vieira, B. Alarcon, J. Ambrus and
L. Olcay, 'Metallic mineralisation in the Gerlache Strait region, Antarctica',
in C. Craddock (ed.), *Antarctic Geoscience*, University of Wisconsin Press,
Madison, 1982, pp. 871–6.

19 See D. D. Hawkes, 'Nature and distribution of metalliferous
mineralisation in the northern Antarctic Peninsula', *Journal of the Geological
Society of London*, CXXXIX, 1982, pp. 803–9; and Vieira, Alarcon, Ambrus
and Olcay, 'Metallic mineralisation in the Gerlache Strait region, Antarctica'.

20 See W. R. Vennum, 'Unusual magnesium- and iron-bearing salts
from West Antarctica', *Antarctic Journal of the US*, XXI, 1986, pp. 55–7; and
W. R. Vennum and J. M. Nishi, 'New Antarctic mineral occurrences',
Antarctic Journal of the US, XVI, 1981, pp. 14–15.

21 M. J. de Wit, *Minerals and Mining in Antarctica – Science and Technology,
Economics and Politics*, Clarendon Press, Oxford, 1986.

22 R. C. R. Willan, 'Review of minerals and mining in Antarctica –
science and technology, economics and politics', *Geological Journal*, XXIV,
1989, pp. 229–31.

23 B. L. Gulson and P. M. Porritt, 'Base metal exploration of the Mount
Read volcanics, western Tasmania; Part II. Lead isotope signatures and
genetic implications', *Economic Geology*, LXXXII, 1987, pp. 291–307.

24 See R. C. R. Willan, 'Metalliferous mineralization in the Antarctic
Peninsula'.

25 R. D. McIver, 'Hydrocarbon gases in canned core samples from Leg
28 Sites 271, 272 and 273', in D. E. Hayes and L. A. Frakes, *et al.*, *Initial
Reports of the Deep Sea Drilling Project*, 28, US Government Printing Office,
Washington, DC, 1975, pp. 815–17.

26 M. J. Whiticar, E. Suess and H. Wehner, 'Thermogenic hydrocarbons in surface sediments of the Bransfield Strait, Antarctic Peninsula',
Nature, CCCXIV, 1985, pp. 87–90.

27 J. C. Behrendt, 'Petroleum and mineral resources of Antarctica', *US
Geological Survey Circular, 909*, Reston, Virginia, 1983.

28 B. St John, 'Antarctica – geology and hydrocarbon potential', in
M. T. Halbouty (ed.), *Future Petroleum Provinces of the World*, American
Association of Petroleum Geologists Memoir, 40, 1986, pp. 55–100.

29 D. A. White and H. M. Gehman, 'Methods of estimating oil and gas

resources', *Bulletin of the American Association of Petroleum Geologists*, LXIII, 1979, pp. 2183–92.

30   D. I. M. Macdonald, P. F. Barker, S. W. Garrett, J. R. Ineson, D. Pirrie, B. C. Storey, A. G. Whitham, R. R. F. Kinghorn and J. E. A. Marshall, 'A preliminary assessment of the hydrocarbon potential of the Larsen Basin, Antarctica', *Marine and Petroleum Geology*, V, 1988, pp. 34–53.

31   E. H. Shackleton, *South*, Heinemann, London, 1919.

32   G. Osherenko and O. R. Young, *The Age of the Arctic*, Cambridge University Press, 1989.

33   St John, 'Antarctica – geology and hydrocarbon potential'.

34   D. J. Drewry, *Antarctica: Glaciological and Geophysical Folio*, Scott Polar Research Institute, Cambridge, 1933.

35   S. K. Donovan, 'The fit of the continents in the late Precambrian', *Nature*, CCCXXVII, 1987, pp. 139–41.

36   M. J. de Wit, M. Jeffery, H. Bergh and L. Nicolaysen, *Geological map of sectors of Gondwana reconstructed to their disposition ~ 150 million years ago*, American Association of Petroleum Geologists and the University of Witwatersrand, Tulsa, Oklahoma, 1988.

The political case for the Minerals Convention

This chapter does not purport to represent an official position. The views expressed are the author's and are not to be taken as necessarily representing the views of the British Government.

Although there is a long tradition of agreement by consensus between the Antarctic Treaty parties, differences of view have emerged about the Antarctic Minerals Convention and the broader issue of environmental protection of Antarctica. This situation warrants consideration of the perceived benefits and problems of the Minerals Convention and of the issues that have been raised in the debate about whether or not it should enter into force. As is apparent from later chapters, there are differing views about the way, and extent to which, Antarctica should contribute to the global good. There are both proponents and opponents of the Minerals Convention.

If such opposing positions give rise to international tension, it is the function of diplomacy to see if they can be resolved. Towards that end it is important to be clear about the nature of the differences giving rise to the tensions. But first we should try to be clear as to where there is common ground between the two sides.

There is, clearly, a consensus that the use of Antarctica as a scientific laboratory is not only acceptable, but is to be encouraged for the good of all. There is also a consensus that nothing done in the Antarctic should have global environmental consequences or so affect the Antarctic environment as to prejudice its use or value for other purposes. As a consequence of these elements of common ground, there is a general belief that any use of Antarctica should have the least possible environmental impact consistent with the achievement of the objective in view. That is not to be read as implying that the ends necessarily justify the means. But, with three exceptions, the Antarctic

Treaty Consultative Parties have proceeded on the tacit understanding that no category of activity should be outlawed unless there
are wholly compelling reasons for doing so.

The exceptions concern the use of Antarctica for military purposes,
the prohibition of nuclear explosions, and the disposal of radioactive
waste. These prohibitions command universal acceptance; the question
at issue is whether there are reasons of a similarly compelling nature to
ban all types of mineral resource activity in Antarctica.

A simple way of describing the differences between those countries
which are seeking a comprehensive environmental protection convention, including a ban on all mining, and those which favour
proceeding with the Antarctic Minerals Convention, is that one side
wishes to outlaw all Antarctic mineral resource activity – that is, to
add Antarctic mineral resource activity to military activity, nuclear
explosions and the dumping of radioactive waste as an activity which
should be outlawed in Antarctica. Those taking that position assert
that there is a prima-facie case that any Antarctic mineral resource
activity would have global environmental consequences, would
prejudice the capacity of Antarctica to continue as a scientific
laboratory, and would prejudice other uses of Antarctica.

The position of the other side is that, while accepting that Antarctic
mineral resource activities could have global consequences, could
disturb the use of Antarctica as a scientific laboratory, and could
prejudice other uses in areas where such mineral activity took place,
the necessary judgements should be made, and precautions be taken,
on a case by case basis. That is to say that the prima-facie case may not
stand up in all circumstances.

It is necessary to go into a little more detail and to determine the
extent of any further elements of common ground between the two
sides and, by so doing, narrow down the area of difference.

First of all there was amongst the thirty-three countries that
participated in the Antarctic Treaty Consultative Party meeting at
Wellington that closed on 2 June 1988, a consensus that exploration
for, and development of, Antarctic mineral resources should be
banned until there is a consensus that exploration for a particular
mineral might proceed. For such a ban to be lifted requires that
consensus judgements be made as to the acceptability of the
environmental consequences of the proposed activity and about
whether enough is known to predict what those environmental
consequences might be. The Convention on the Regulation of

Antarctic Mineral Resource Activities (CRAMRA) was the international instrument that had been negotiated to ensure that such a conditional ban on exploration and development of Antarctic minerals could be implemented.

Those who have taken the position that CRAMRA should not enter into force also favour a ban on exploration and development. To that extent, there is an element of common ground between the two sides. The difference, of course, is that whereas the CRAMRA ban is conditional on a consensus to lift it, the ban being sought by those not favouring CRAMRA is absolute. They seek for it the same authority as the bans on military activity, nuclear explosions and the dumping of radioactive waste. It can, however, be argued that the difference, despite its crucial appearance, is not quite as great as it seems.

As a matter of international law, an international legal instrument, such as a treaty or a convention, can be changed in any way, or even cancelled, if there is a consensus amongst the contracting parties to do so. It follows from this that however absolute is the wording contained in a treaty banning exploration and development, it could be changed by a consensus amongst the parties to it. It was suggested at a meeting of the Australian Institute of International Affairs in Hobart in November 1989, that the fact that the CRAMRA ban and the absolute ban could both be lifted by consensus, puts the two bans on an equal footing as to their effectiveness. While that point is arguable, there are nevertheless grounds, in international law at any rate, for looking at the two bans, 'conditional' and 'absolute', as having similar force.

There is, however, another, and probably rather more important difference between those supporting CRAMRA and those seeking an absolute ban. The difference relates to prospecting – the stage of activity which comes before exploration and development. CRAMRA provides that prospecting can go ahead in the Antarctic if the country where it is organised – the 'sponsoring state' in CRAMRA parlance – is satisfied that the particular proposed prospecting activity would be consistent with the requirements of the Convention.[1] Those seeking an absolute ban, wish to include under it a ban on prospecting. This is a crucial difference and one which merits close attention.

A useful starting point is to examine the limited nature of activity that can proceed under the heading of prospecting. In accordance with Article 1(8) of the Convention:

' "Prospecting" means activities, including logistic support, aimed at identifying areas of mineral resource potential for possible exploration and development, including geological, geochemical and geophysical investigations and field observations, the use of remote sensing techniques and collection of surface, seafloor and sub-ice samples. Such activities do not include dredging and excavations, except for the purpose of obtaining small-scale samples, or drilling, except shallow drilling into rock and sediment to depths not exceeding twenty-five metres, or such other depth as the Commission may determine for particular circumstances."[2]

The negotiators of CRAMRA were aware that prospecting was a term which, elsewhere in the world, was commonly used to include activities which could have serious environmental implications. To follow such a practice in the Antarctic was looked upon as being environmentally unacceptable. Consequently prospecting had to be distinguished from the later stages by means of a definition which made environmental sense. In adopting such a restrictive definition, the CRAMRA negotiators were very much aware that their definition was virtually identical to a definition of geological and solid earth geophysical research carried out for scientific purposes. This was done on purpose. The vital difference, however, is that prospecting under CRAMRA would be subject to the more stringent controls and regulations provided for in the Convention.

The reasoning of those seeking a ban which would cover prospecting as well as all later stages of activity is, essentially, based on the 'thin end of the wedge' argument. That is to say, that once prospecting has been allowed, exploration and development are sure to follow. There is nothing in the Convention to suggest that this is the case – indeed, the structure of the Convention, by putting the need for the consensus decision **after** prospecting and **before** exploration, is intended to counter any thoughts of exploration inevitably following prospecting. The pure logic of the proposition is, in any case, faulty inasmuch as any inevitability must depend on the results of prospecting activity being positive. If they are negative then exploration and development will not follow. Leaving that logical point aside, there is a certain degree of common ground between both sides in that those supporting CRAMRA and those seeking the all-embracing ban, agree that in such a vast place as Antarctica, the possibility that there could be mineral deposits which might be of value cannot be ruled out.

The issue then is whether a ban on Antarctic prospecting could be made to work. It is worth reiterating here that all that could be done under the prospecting provisions of CRAMRA, could also be done under the freedom of scientific research provisions of the Antarctic Treaty. Because of the limited nature of what is allowed under prospecting and its similarity with scientific research, it can be argued that the person much beloved of English jurists, the 'reasonable man' on the Clapham omnibus, would conclude that any assurances that a ban on prospecting would work were likely to turn out to be insubstantial.

Apart from the tighter controls that would apply to prospecting, the only other difference between prospecting under CRAMRA and scientific research of a similar nature under the Treaty relates, not to what is done on the ground, but to what happens to the end product. Under CRAMRA, the results of prospecting can be kept confidential for a certain period. Under the Antarctic Treaty the results of scientific research shall, 'to the greatest extent feasible and practicable' be exchanged between Contracting Parties.[5]

There are those who have cast doubt on whether the freedom of information provisions of the Treaty are being complied with in respect of geological and geophysical research. In order to refute such allegations it is necessary to find some means of ensuring that the scientists concerned have completely, and without any reservation, emptied their heads as well as their filing cabinets. Such a proposition can be regarded as an impossibility. It is for that reason that such allegations create fertile ground for corrosive suspicion. There is, however, a point of broader significance to make. That is, that unless Article II of the Antarctic Treaty is amended so that geological and geophysical research are specifically excluded from the Treaty's provision for the freedom of scientific investigation, there is no way in which knowledge of the mineral potential of Antarctica will not steadily accrue. Such knowledge could be positive or negative in terms of mineral discoveries.

The position of those seeking an all-embracing ban is that the likelihood is that some of it will be positive. If they took the view that it was all going to be negative, the basis for the ban they are seeking would evaporate. The position of those seeking the entry into force of CRAMRA is that some of that increase in knowledge would be negative and some would be positive. Taken together, they would tend, slowly but surely, to identify areas of greater prospectivity.

Returning to the possibility that either a CRAMRA ban or an absolute ban could be lifted by consensus, and the suggestion that both bans were on an equal footing, it is necessary to examine why that may not be the case. While there is a certain equivalence in international law in so far as both bans need consensus to lift them, they are clearly not politically equivalent. That is because while it is possible to foresee the conditional CRAMRA ban being lifted if the relevant conditions can be met, it is very much more difficult to foresee a consensus in favour of lifting an absolute ban.

Looking to the future, it is the political consequences of an absolute ban which are of concern particularly in light of the fact that knowledge of where minerals are likely to be found will steadily accrue unless the Antarctic Treaty is amended to prevent it. Such an amendment could not be negotiated. It follows that a steady increase in knowledge about mineral resources is inevitable. It does not much matter whether such knowledge is kept secret or is made public. Its effects will be the same in so far as interests will become, covertly or overtly, increasingly focused on particular minerals and particular areas. Even if on the face of it results are published, there will always be some doubt that not everything has been made public. As an unhappy corollary of geological and geophysical research, suspicion between parties to the Treaty will tend to grow. In the event of positive minerals discoveries the interests of the relevant claimant state and the state whose scientists made the positive observations will become vested. An atmosphere of competition and mutual suspicion about each country's discoveries would develop.

Only two per cent of Antarctica's rock surface has yet been looked at by anyone who would remotely qualify as an economic geologist. The potential for finding something of value remains. It may well be that even if the other ninety-eight per cent is looked at, nothing will be found. But if something is found, what then?

There are, it seems, three possibilities. First, that the absolute ban would hold and that the finding state and the relevant claimant state (possibly one and the same) would shrug their shoulders and walk away from it; secondly, that some sort of international agreement amongst the Consultative Parties would be achieved – CRAMRA revisited; or, thirdly, that there would be an attempt by the states immediately concerned to come to a cosy deal.

The absolute ban would hold only in circumstances where the need for the deposit that had been tentatively located was not sufficient to warrant a challenge to the ban.

The second possibility depends on an inherently unlikely situation that the states with vested interests would be prepared to share them with all other parties. It follows that if the deposit is of sufficient economic value, the third option would be the most likely. But by that time a degree of frustration and suspicion would have built up within the Antarctic Treaty system. Instead of being the preferred vehicle for the expression of national interest, it would come to be seen as a drag on national interest. Claimants and non-claimants would be at loggerheads and the virtues of Article IV, which provides for the 'freezing' of territorial claims, would come increasingly to be questioned. It must be doubtful if the Antarctic Treaty system could sustain such a challenge.

Whether the mineral deposit that had caused the trouble would then be exploited remains open to considerable doubt. To those who see the continued absence of mineral activity from the Antarctic as being the ultimately desirable objective, such an outcome might appear satisfactory. But it does seem reasonable to question whether the most prudent path to follow is one that has the potential, by undermining the Antarctic Treaty system, to threaten the maintenance of international harmony in the Antarctic.

Those who favour the absolute ban argue in aid of their case that there is a growing international consensus in their favour. In one sense there does indeed appear to be a Calvinistically inclined mood in some parts of the world that there should be at least one place on earth that mankind undertakes not to despoil with industrial development. While one can agree with that mood, it is ethically rather than objectively based.

There is a perennial argument as to whether international affairs should be based on moral absolutes or national interests. It is an argument upon which there is no consensus; nor is there likely to be in the forseeable future. There are certain facets of international relations which must be governed by what is morally and ethically right. The banning of military activity and nuclear explosions in the Antarctic is a response to just such a moral imperative. So also, perhaps, may be the ban on the disposal of radioactive waste in Antarctica. But when it comes to an absolute ban on all Antarctic mineral resource activities, the arguments about the ethical case are more dubious.

As Arthur Schlesinger Jr., recently wrote, [4] there are two questions which can be asked of a policy, 'is it morally right' and 'will it work'. It seems that the absolute ban policy is open to legitimate doubts as to whether it will work. Also its moral or ethical basis is open to question as to who gains and who loses. The CRAMRA policy avoids the 'will it work' pitfalls of the absolute ban policy and leaves the ethical question to be answered in the light of the circumstances at the time when it comes, if ever, to be tested.

At the beginning of this chapter it was suggested that it was the task of diplomacy to see if means could be found of resolving international tensions where they are seen to exist. Let us look for a moment at the probable turn of future events in Antarctica. As things stand at present, if an absolute ban is to become effective, the fifteen Consultative Parties amongst the nineteen States which have signed CRAMRA need to revoke the undertaking they have implicitly given by signing the Convention that they intend to go on to ratify it. If, on the other hand, CRAMRA is to enter into force, a smaller group of countries will need to change the position they are currently occupying.

If neither side is prepared to move and each side demands total capitulation on the part of the other, we shall be faced with the prospect of a prolonged stalemate and a situation in which the conditional gentleman's 'moratorium' on mineral activity, 'pending the timely entry into force' of CRAMRA,' will come under increasing strain and perhaps begin to crumble. Activity which is seen by some as scientific research, and by others as prospecting in disguise, could be the corrosive catalyst.

Such a situation would not be in the real interests of either side and points strongly towards the need for a bridge to be built upon which both sides can meet with their honour more or less intact. There would seem to be four conditions which appear to be necessary for progress to be made The first condition is that neither side should claim any sort of moral superiority over the other (it is too easy in such circumstances for each side to denigrate the motives of the other); the second is that those supporting CRAMRA should be prepared to accept some fettering of the freedoms they would have under the Convention as it stands; the third is that those supporting an absolute ban should be prepared to accept that their interest in ensuring the maintenance of international harmony in Antarctica should balance and modify their interest in the absolute nature of the ban they are seeking, and the fourth is that both sides be prepared to undertake, in good faith, the negotiation of comprehensive measures for the protection of the Antarctic environment and of the liability protocol called for under CRAMRA.

For almost thirty years the development of the Antarctic Treaty system has depended on practising the art of compromise in which both sides in an argument recognise that because of the political need to make decisions by consensus, neither side can get all they want. There does not appear to be any overriding reason why the argument about a conditional ban or an absolute ban on minerals activities should not be subject to the same practice.

Notes

1   See CRAMRA, Article 37.

2   See Antarctic Treaty, Article I(8).

3   See Antarctic Treaty, Article III.

4   Arthur M Schlesinger, Jr., *The Cycles of American History*, André Deutsch Ltd, London, 1986.

5   See 'Antarctic Treaty Recommendations, IX–1: Antarctic Mineral Resources', and 'Final Act of the Fourth Special Antarctic Treaty Consultative Meeting on Antarctic Mineral Resources', in *Handbook of the Antarctic Treaty System*, Sixth Edition, Polar Publications, Cambridge, 1988, pp. 3304–05 and 3309–10 respectively.

Comprehensive environmental protection of the Antarctic: new approaches for new times

Introduction: common ends and many means

All countries with an active interest in Antarctica are intent on protecting the Antarctic environment. They are also intent on preserving and strengthening the Antarctic Treaty system. Yet within this framework of common objectives there is a divergency of views on what should be done. The Australian Government has been closely associated with this debate. It is relevant, therefore, to set out what the Australian Government is seeking to achieve and why.

The Australian objective can be described quite succinctly. On 22 May 1989, the Australian Government announced that it 'is dedicated to the comprehensive protection of the Antarctic environment and in that context our strong commitment is that no mining at all – including oil drilling – should take place in and around the continent.' It added that 'Australia will not sign the Minerals Convention, but instead will pursue the urgent negotiation of a comprehensive environmental protection convention within the framework of the Antarctic Treaty system. In that context, Australia will specifically explore the prospects for the establishment of an 'Antarctic Wilderness Park.'' The terminology has since changed a little, after consultations with the French Government, which takes a similar position, and following discussion at the October 1989 Fifteenth Consultative Meeting in Paris. Recommendation XV-1 of that meeting, which establishes a Special Consultative Meeting to consider comprehensive environmental measures, describes the Australian and French proposal as for 'a comprehensive Convention for the protection of the Antarctic environment which would establish Antarctica as a natural reserve, land of science'. The essence of the proposal is a legally binding and comprehensive regime for environmental protection and a ban on mining. Australia believes that mining in the Antarctic would be grossly incompatible with proper environmental protection.

The new factors

Concern which greeted the Australian announcement – and for that matter the earlier public declaration of the French Prime Minister, M. Rocard, on 20 April 1989 that 'France will not ratify the Antarctic [Minerals Convention] as it stands' – was as much born of concern about the way of doing things as the substance of the announcements. The announcements seemed to represent an abandonment of the course that the Consultative Parties had set themselves for the previous six years or more and injected a range of new political considerations into Antarctic Treaty decision making.

International environmental concern

These new political considerations have their inspiration very much in increased international concern about the environment. It is a concern, which in Australia and the United Kingdom, as much as anywhere, is a grass roots factor that is influencing governments from below. The rise in activism and influence of non-governmental environmental organisations is both a token of grass roots concern and a potent factor in mobilising it. In the United States, for example, environmental groups have a membership growing faster than any other category of organisation. United States' groups reportedly now count their membership in millions.

It is thus no coincidence that the Australian and French decisions on Antarctica bear many similarities with the Antarctic policies which environmental groups have been espousing for much of this decade. Greenpeace and the Australian Conservation Foundation particularly come to mind in the Australian context. In France the role of non-governmental organisations is symbolised by the June 1989 report on Antarctica of the Cousteau Foundation given at the request of the French President. A point to emphasise is that in Australia at least, there was a high level of public concern about Antarctica manifested in letters to ministers, petitions and public debate which had no precedent in earlier consideration in Australia of Antarctic issues.

Such grass roots concern obviously has effect at the political level. In Australia, before the decision on 22 May 1989, this was manifested by the decision of the Federal Parliamentary Opposition on 2 May calling for a convention to ban mining in the Antarctic and a Senate[2] motion to similar effect adopted the following day. Indeed the familiar pattern of a non-government majority in the Senate makes it extremely unlikely that any future Australian government will change its position on Antarctica even if it were inclined to. That majority would almost certainly block passage in the Senate of any legislation seeking to ratify the Minerals Convention.

Parliamentary manifestations of political interest in Antarctic conservation are evident in a range of countries. In the United Kingdom, the Labour Party broke with a long standing tradition of bipartisan support for Antarctic issues by voting in July 1989 in the House of Commons against legislation to implement the Minerals Convention. In France[3] and Sweden[4] members have asked parliamentary questions sympathetic to a conservation approach. There is a preliminary French parliamentary report on the subject of Antarctic mining.[5]

The European Parliament on 18 September 1987 passed two resolutions on Antarctica: one on the protection of the environment and wildlife in Antarctica, and the other on the economic significance of Antarctica and the Antarctic Ocean.

Even the latter included many environmental points including the statement that 'any exploration or exploitation of minerals would adversely affect the unique values of the Antarctic environment.'

In the Federal Republic of Germany there were debates in the Bundestag on 5 May 1988 and 1 June 1989 concerning Antarctic conservation and the Minerals Convention. The Belgian Parliament held a debate which led on 30 June 1989 to the passage of legislation to prohibit Belgian nationals and corporations from undertaking minerals activities in the Antarctic. The Chamber of Deputies of the Italian Parliament carried a motion unanimously on 28 September 1989 'committing' the Italian Government 'not to subscribe to and not to ratify the Minerals Convention and to support and promote the proposal to transform Antarctica into the first world natural park among the Consultative Parties of the Antarctic Treaty.'

In New Zealand the Government announced on 26 February 1990 that it will not now move to ratify the minerals Convention and declared that the Antarctic community should 'move constructively in the direction of the Australian and New Zealand proposals on environment protection measures.'[6] The Opposition had earlier announced that if it came to power it would not sign the Minerals Convention but would work with other governments to ban mining and ensure the protection of the region.[7]

On 26 September 1989 Senator Gore and nine other senators submitted a Joint Resolution calling for the United States to encourage immediate negotiations towards a new agreement among Antarctic Treaty Consultative Parties, for the full protection of Antarctica as a global ecological commons.[8] On 7 February 1990, Representative Conte introduced a bill which would prevent prospecting for, and the exploration and exploitation of, Antarctic mineral resources by United States citizens, regulate tourist expeditions and direct the Secretary of State to enter into negotiations with other Antarctic Treaty Parties to ban mineral resource activities and permanently conserve and protect the Antarctic environment.

New dynamics in Antarctic decision-making

Whatever the merits of this political level focus on Antarctic environmental issues, it nevertheless represents a new dimension which the Antarctic Treaty Parties have to take into account. Science has supplied the insights about how the world works and the place of Antarctica in it. However, concern about the Antarctic has spread far beyond the scientific community. Indeed, one more reason for the Australian Government's May 1989 decision being the watershed it was is that it bypassed that community which had been so influential in previous important Antarctic decisions. Scientists in Australia should not consider themselves discriminated against. Many public servants felt equally marginalised. These observations simply emphasise the point that we are now in a new ball game.

The possible responses for the protection of the Antarctic environment

While the Australian Government's decision was radical in terms of the course of action on which the Antarctic Treaty Parties had been embarked, namely conclusion of a Minerals Convention, it was made quite clear that Australia would pursue the negotiation of a comprehensive environmental protection regime 'within the framework of the Antarctic Treaty system'. The announcement even acknowledged that 'the recently concluded Convention on the

Regulation of Antarctic Mineral Resource Activities (CRAMRA) is very much better than no protective regime of any kind in relation to these activities.' The Government nevertheless expressed the belief that 'it is both desirable and possible to seek stronger protection for what remains the world's last great wilderness.'

The Antarctic Treaty system with its strategic and other assurances is too important for Australia to jeopardise. It also makes good environmental sense to enhance Antarctic environmental protection through the Antarctic Treaty system. The system includes all countries active in the Antarctic. Those countries have, through the system, also set in place a series of environmental measures and instruments which make the Antarctic probably the best example of regional environmental co-operation in the world.

Australia and France pretend no monopoly of wisdom on what a comprehensive environmental regime for Antarctica should ultimately contain.

New Zealand has said that: '. . . in order to ensure a properly integrated, comprehensive and internally consistent regime of environmental protection, it is desirable that there be elaborated a series of basic standards that must be met by all forms of human activity in the Treaty area.'[9] Sweden is on record that 'basic standards for all forms of human activities in the Treaty Area should be elaborated. All activities in Antarctica should be submitted to effective environmental standards.'[10] Australia supports these principles.

Provision largely exists under the present system of Consultative Meetings for environmental measures to be developed. Australia believes that with a minimum of change this process can be adapted to permit the establishment of detailed measures to achieve the general standards and to provide for the effective implementation of the regime. This would enable specific rules to be developed to cover aspects of activities such as tourism where this is shown to be necessary.

Chile has stated that: 'All activities, including scientific research, tourism, navigation and the installation of new stations should be preceded [by] evaluation of their environmental impact . . .'.[11] Chile also has proposed that there should be adopted 'a set of rules of a mandatory nature' set out in a 'binding instrument containing the Comprehensive Measures'.[12] Australia agrees with these suggestions also.

A protected area system already exists in Antarctica. As New Zealand has pointed out: 'Establishment of areas within Antarctica to

be given special protection is an effective way of ensuring the preservation of unique wildlife and vegetation sites and other outstanding
natural, cultural, historic and scientific features.'[13] Australia agrees with
this assessement and also agrees with suggestions that the existing
system could be rationalised by establishing simpler categories. Areas
and uses designated under the existing system should be recognised
and given an equivalent level of protection and management.

Sweden and Chile have stated that questions of liability should be
addressed.[14] Australia agrees that the issue of liability for environmental harm is an important one.

The United States has expressed the belief that the Parties to the
Antarctic Treaty should 'ensure the conduct of monitoring necessary
to assess the effectiveness of the environmental protection provisions
and to identify the need for strengthening and supplementing them'.[15]
Australia is willing to play its part in this area.

The current system of national inspection has come to include
inspection of compliance with existing environmental measures. As
New Zealand has pointed out that '. . . because of the growing
recognition of the importance of the protection of the Antarctic
environment and the need for the Treaty Parties to establish an
effective environmental protection regime' it may also be appropriate
to adopt the precedent offered by the Minerals Convention and
develop a system of institutional inspection.[16] Australia also supports
the thrust of this approach.

A comprehensive approach

Discussion at the Fifteenth Consultative Meeting thus revealed much
common ground on what an enhanced environmental protection
regime should contain.

There was also general recognition that the Antarctic Treaty parties
should continue and even redouble their efforts to protect the
Antarctic environment. There were broadly two different views on the
best way to do it: to follow the pattern that has been set for the last
thirty years and adopt a series of recommendations to cover
environmental problems as they arise for particular activities, or
attempt to formulate a series of rules applicable to all human activity.

In a sense the Minerals Convention was an example of the existing
approach. While not a recommendation, it covered one particular
group of activities – prospecting, exploration and mining. Those who

grappled with the problems of developing that regime were conscious that many of the environmental concerns for which rules were being devised in that context were common to human activities other than mining. Given that it is reputedly an Anglo-Saxon predilection to favour moving from the particular towards the general it might have been expected that Australia would favour intensification of the present approach. Whether under the influence of Australia's French colleagues or not, Australia has in fact opted for the more conceptual approach of establishment of general rules – call it codification – covering all human activities.

In a way, virtually all the Consultative Parties at the 1989 Paris meeting conceded the strongest intellectual argument in favour of a comprehensive environmental regime. That is, there needs to be more integration and greater co-ordination than the present system provides. In the words of New Zealand: 'Clearly, the existing arrangements taken as a whole are not yet sufficient in scope or implementation to provide the protection that Antarctica warrants, but they do provide a solid foundation on which to build.'[17] In seeking to achieve protection of the environment, there is no justification for environmental impact evaluation procedures applying to one form of activity but not to another, the intensity of the impact of which is likely to be as great. (At present procedures apply only to scientific activities.) The more one tries to address particular problems the more one finds that they are interrelated.

The widely accepted objectives of integrated rules, legal enforceability and comprehensiveness, point logically to a unified regime of treaty status. The contemplation of that possibility now is hardly a radical proposal given the evolution of environmental measures within the Treaty system to date. It is also worth bearing in mind that as well as serving a practical need, a comprehensive environmental regime would fill a gap in the Antarctic Treaty itself. It is a measure of how much things have changed that this Treaty contains no reference to preservation of the environment in its guiding principles set out in the preamble, or in the key substantive articles. Indeed the word 'environment' is not used at all in the Treaty. The only environmental reference is to the 'preservation and conservation of living resources in Antarctica'.[18] As if to remedy this, as long ago as 1961 and 1962, Chile and the United Kingdom proposed that a convention would be an appropriate form for an instrument on the conservation of wildlife in the Antarctic.[19]

Where does the Minerals Convention fit?

The essential requirement for environmental protection of the Antarctic is a comprehensive environmental protection regime of some sort. In this sense the question of the Minerals Convention, which has been the focus of public debate, is a side issue. The two became associated because they were linked in the public political debate in Australia and other countries. Environmental groups which had long argued for a comprehensive environmental approach were now urging this course upon governments in the particular context of their opposition to adherence to the Minerals Convention. The criticism that the Minerals Convention attracted for quite different reasons, as much from these public debates as from governments critical of the Antarctic Treaty system, has shown that Convention to be a public relations disaster.

In a substantive sense the Minerals Convention is a largely separate issue. For that reason the Australian Government has maintained that the question of the Minerals Convention should be set aside while consideration is being given to a comprehensive environmental regime. Nevertheless there are at least two important links which make it necessary for there to be at least some discussion of mineral resource activities in the context of a comprehensive approach to the protection of the environment. In the first place, a link follows from the assessment that many have made that mining and related activities would not be consistent with the adequate protection of the Antarctic environment and should be prohibited. Secondly, some defenders of the Minerals Convention have asserted that it would be necessary to exclude the subject matter of the Minerals Convention from a general environmental regime, because the Minerals Convention can be relied upon to protect the Antarctic from unacceptable environmental damage and that it amounts to a prohibition on mining. The essential concern about mining is not the merits or otherwise of the Minerals Convention, but whether the possibility of Antarctic mining should be closed off for the sake of the environmental protection of the Antarctic environment and the other uses to which Antarctica may be put.

Scientists on mining

When Australia took the position that mining should be prohibited and not just 'regulated' it was greeted with accusations of jumping to arbitrary conclusions which had no scientific basis. This accusation is familiar in one form or another in most environmental debates: where is the scientific proof to justify the control? It is notoriously hard to develop proof of environmental links satisfying strict scientific criteria at any stage before hypothetical adverse impacts become a reality. On the other hand, there is a great deal of scientific opinion which is strongly suggestive of prohibition being the only prudent course. If there is to be a burden of proof, it should lie on those seeking to show that there is a reasonable likelihood that mining could be safe.

A survey Australia has made of the consideration from the middle of the 1970s of the environmental implications of Antarctic mining reveals many serious scientific reservations about the compatibility of mining with environmental protection. The consideration concerned was undertaken by the Scientific Committee on Antarctic Research (SCAR) and other expert groups at the request of the Consultative Parties. These reservations are all the more striking given the ostensible care of the experts involved to avoid value judgements on whether Antarctic mining should proceed. However, this is not the place to give a detailed account of this early work.

Some experts say that: 'It is likely that if development occurs, it will initially take place either in the Ross Sea embayment or adjacent to the Antarctic Peninsula.'[20] Accessible areas of the Antarctic Peninsula as well as offshore sedimentary basins are particularly vulnerable areas: 'In Antarctica, sedimentary deposits are generally found on the relatively narrow continental shelves and arc extensions and in the Ross, Weddell and smaller basins around the continent. It is in these regions that there is high biological productivity'.[21] There is thus likely to be little consolation from the allegedly large buffering effect of the vast Southern Ocean. The areas most at risk from hydrocarbon and other pollution are likely to be the areas subject to pack-ice or close to land. A SCAR report noted that: 'The buffering capacity of this ocean is correspondingly great and open ocean impacts would be slight. In contrast, the potential impacts in continental shelf seas and around islands would be very much greater.'[22]

Whether onshore or offshore mining is involved, 'Massive terrain modification is unavoidable in the construction of shore installations, quarters, and port and industrial facilities for handling oil or ore products on a commercial level. Major impacts would be on the terrestrial permafrost regime, surface and subsurface drainage patterns, and the "destruction and/or modification of the soils and terrestrial biota".' The extreme climate of the region is likely to increase substantially the risks of equipment failure and human error even taking into account improvements in technology and standards. 'Tanker accidents due to collisions with icebergs would be expected to be more frequent than in other seas, and routine operational discharges worse than elsewhere owing to the effect of the greater rigours of the environment in enhancing human errors'.[4]

Given these sort of perils one might question why the Antarctic Treaty Parties have invested so much effort in the development of environmental protection if they are willing to tolerate the risks inherent in mining operations. What is the point of regulating waste disposal around a small scientific station or conducting an environmental impact evaluation of a series of scientific experiments when those same parties are prepared to contemplate the much more severe impact inherent in setting up a mining venture? It is appropriate to recall a view put at a working group in 1973 when the possibility of a mining regime was first being thought about:

'Commercial mineral exploration and [mining] would disturb scientific activity, would constitute a serious danger for the Antarctic ecosystem, and could cause the contamination of its seas, ice and land despite the fact that the requirements of scientific investigation are that the Antarctic should be a non-contaminated area. The attention of the Group was drawn to the preoccupation of successive Consultative Meetings with the preservation of fauna and flora, which, according to one view, was incompatible with commercial mineral exploration."[5]

Is it possible to reconcile mining and environmental protection?

The answer given to the question posed above is that the Minerals Convention prevents mining occurring unless it is shown to be safe. On at least one level this is correct but Australia fears that the solution is too neat. The Convention has several objectives. As its title indicates, one of these was to regulate any possible mineral resource activity in the Antarctic. Another aim was that 'the protection of the Antarctic environment must be a basic consideration' in any decision to carry out such activity. It was believed possible to reconcile these objectives.

The Convention says as much. The parties were: 'Concerned to ensure that Antarctic mineral resource activities, should they occur, are compatible with scientific investigation in Antarctica and other legitimate uses of Antarctica.'[26]

There is little evidence that an examination was ever conducted of whether it would ever be reasonably likely that the twin objectives would be reconcilable. The expert reports on possible minerals activity all seemed to avoid confronting this point for the good reason that the groups were not asked to address it. The expert group that came closest to doing so was one of 1977, which in an admittedly rushed fashion, sought to evaluate 'the sensitivity of Antarctic marine and terrestrial ecosystems to different forms of environmental disturbance and the extent to which certain marine or terrestrial environments might require special protection.'[27] While it concluded that a number of ecosystems would indeed be very sensitive to mining it did not go on to examine whether mining could ever be safe in accordance with a range of possible standards.

Would a mining ban be inherently unstable?

Some have said that a ban on mining would be inherently unstable. This argument is not convincing given the demonstrable stability of other outright bans included in the Antarctic Treaty itself concerning militarisation of the Antarctic, nuclear explosions and deposit of nuclear waste there. Rather a regime which, to use an Australian phrase, 'has a bob both ways', would be inherently far less stable. With mining in prospect – and encouraged by the fairly free range for prospecting and promise of security of title at the end of the day – the pressures on the decision-making structure of the Minerals Convention could well be enormous. This is not an encouraging prospect given that, as a recent US Congressional study points out, 'Ultimately, the effectiveness of environmental protection under the Minerals Convention rests largely with the political will of the Parties.'[28]

The Minerals Convention is Not a Ban

It is said that the Minerals Convention imposes a ban on mining. This follows from the consensus that is required in order to open an area for possible exploration and development. However, the relevant provision is not phrased in terms of a ban. The requirement of unanimity is just a procedural rule determining how decisions may be made. One has to look at the decision that is provided for. If unanimity amounts to a ban one could just as well say that there was a ban on the Consultative Parties under the Antarctic Treaty approving measures under Article IX. The Minerals Convention makes clear that the right to refuse consent under Article 41(2) to the opening of an area for exploration or development is not an absolute one but arises only if the objection is 'consistent with the Convention'. A party might be challenged under the dispute settlement procedure if it withheld its consent for environmental or other reasons not sanctioned by the Convention.

Uncertainty of key tests

The prospect of both destabilising conflict, and that a state which takes a pro-environmental position might be ruled not to be acting consistently with the Convention, is made all the more likely by the vagueness of at least one key test. The important environmental standards set out in Article 4 rely largely on the concept of 'significant' harm. That is a threshold test, yet it is not defined in the Convention. There could be all manner of controversy about its meaning but there are a number of reasons for believing that a pro-conservation view of it would not prevail. This is suggested by the virtual inevitability of high impact from mining, the fact that mining is envisaged as a potentially acceptable activity in the objectives of the Convention, and differing value judgements about what is, or is not, environmentally acceptable.

Pragmatic arguments in favour of the Minerals Convention

Although aspects of the Minerals Convention can be criticised it must be acknowledged that it has many merits from an environmental point of view.

These merits are thrown into relief when a series of scenarios are painted showing that it is impossible to get anything better and that things could be a great deal worse without the Minerals Convention. These are pragmatic arguments which have considerable force if one accepts the assessments that underlie them.

The Minerals Convention can indeed prevent unregulated mining at least by nationals of the states which are a party to it. But which are the states whose nationals are most likely to engage in mineral resource activity in terms of access to capital and technology? One would have to say that most, if not all, are a party to the Antarctic Treaty system. If a non-party state is involved there are immediate difficulties in enforcement which a provision like Article 7(5) of the Minerals Convention (itself modelled on Article X of the Antarctic Treaty) acknowledges.

The judgement has also been made that it is unlikely that anything better will ever be put in place. New Zealand's abortive proposal in 1975 for a permanent moratorium on mining is cited as showing that this course was tried and failed. That was fifteen years ago though. The political circumstances outlined earlier suggest the desirability of trying this option again.

It is also said that the course urged by Australia and France would destabilise the Antarctic Treaty system. It will only do so if there is failure at the end of the day to reach agreement. Again, because of the political considerations mentioned above, Australia believes that it will be possible to reach agreement on a regime with stronger environmental safeguards than the Minerals Convention. If indeed that agreement is reached, there can be no doubt but that the Antarctic Treaty system will be greatly strengthened. The last General Assembly of the United Nations shows that the nub of international criticism of the Treaty system would be removed if the Parties to it were to renounce the possibility of mining and of related economic gain to themselves from it.

There is an air of unreality around these various pragmatic arguments. All agree that there is no short- or even medium- term likelihood of mineral resource activity (except prospecting) occurring. In a practical and political context, no Treaty Party is going to break the existing moratorium on those activities. There is, therefore, ample time to negotiate a comprehensive environmental protection regime. To use the words of the Australian Government's announcement of 22 May 1989: 'Although . . . CRAMRA is very much better than no protective regime of any kind in relation to [mineral resource] activities, we believe that it is both desirable and possible to seek stronger protection for what remains the world's last great wilderness.'

Notes

1   R. J. L Hawke, Prime Minister of Australia, *Protection of the Antarctic Environment*, Press Release, 22 May 1989.

2   The Senate is the upper house in the bicameral Australian Federal Parliament.

3   Answer in the Chamber of Deputies on about 10 August 1989 to a question of Pierre Brantus and answer of the Associate Minister for Foreign Affairs (Mme Avice) on 25 October 1989.

4   Answers of 1 December 1988 and 24 October 1989 of the Swedish Foreign Minister (Sten Andersson).

5   Le Deaut, *Etude de faisabilite du rapport sur les problemes poses par le developpement des activites liees a l'extraction des ressources minerales de l'Antarctique*, Assemblee Nationale, Senat, Office parlementaire d'Evaluation des Choix scientifiques et technologiques, Paris, 1989, p. 46.

6   The Rt. Hon. Geoffrey Palmer, Prime Minister of New Zealand, Press Statement, 26 February 1990.

7   J. B. Bolger, Leader of the Opposition, Speech at Wellington Clinical School, 4 August 1989.

8   Congressional Record, Senate, Washington, DC, 26 September 1989, pp. S11, 906–8.

9   'Working paper on the protection of the Antarctic environment', submitted by the New Zealand delegation (document XV ATCM/WP/4), p. 2.

10   Working paper on 'Comprehensive measure[s] for the protection [of the] Antarctic environment and dependent and associated ecosystems: proposed outline of a draft recommendation by the Sweden Delegation' (document XV ATCM/WP/14), para. 3.

11   'Working paper submitted by the Delegation of Chile' (document XV ATCM/WP/7) p. 7.

12   Document XV ATCM/WP/7, n. 11, pp. 1 and 4.

13   Document XV ATCM/WP/4, n. 9, p. 7.

14   Document XV ATCM/WP/14, n. 10, para. 14; and document XV ATCM/WP/7, n. 11, p. 7.

15   'Comprehensive measures for the protection of the Antarctic environment and dependent and associated ecosystem: a working paper submitted by the United States' (document XV ATCM/WP/8) p. 3.

16   Document XV ATCM/WP/4, n. 9, p. 15.

17   *Ibid.* pp. 3–4.

18   Article IX(2) of the Antarctic Treaty.

19   ATCM I document: DOC./P.15 (CORRI) of 13 July 1961 and ATCM II document: P.3 of 17 July 1962.

20   SCAR, *Report number 2 of Scientific Committee on Antarctic Research, Group of Specialists on Antarctic Environmental Implications of Possible Mineral Exploration and Exploitation (AEIMEE)*, Cambridge (roneoed), June 1983, p. 4.

21   Zumberge, J. H. (ed.), *Possible environmental effects of mineral exploration and exploitation in Antarctica: an adaption of a Report by the Group of Specialists on*

the Environmental Impact Assessment of Mineral Resource Exploration and Exploitation in Antarctica (EAMREA) Convened by SCAR in response to Recommendation VII–14 of the Eighth Antarctic Consultative Meeting, SCAR, Cambridge, March 1979, p. 26.

22 SCAR, *AEIMEE report number 2*, footnote 19, p. 22; and SCAR, *Report number I of Scientific Committee on Antarctic Research, Group of Specialists on Antarctic Environmental Implications of Possible Mineral Exploration and Exploitation (AEIMEE)*, University of Nebraska–Lincoln (roneoed), Nebraska, USA, 26–9 May 1981, p. 7.

23 *Preliminary Assessment by a SCAR Group of Specialists on the Environmental Impact of Mineral Exploration/Exploitation in Antarctica*, EAMREA, Typescript, 1977, p. 47.

24 *Ibid.* p. 58.

25 'Report of the working group on legal and political questions of the Report of the meeting of Experts organised by the Fridtjof Nansen Foundation on Existing Law relevant to the authorisation or prohibition of mineral exploration for commercial purposes in the Antarctic Treaty area', in *US policy with respect to mineral exploration and exploitation in the Antarctic*, Subcommittee on Oceans and International Environment of the Senate Committee on Foreign Relations in US Senate, *Congressional record of hearing*, 94th Congress, Ist session, 15 May 1975, pp. 76–80 at para. 22.

26 Preamble, Convention on the Regulation of Antarctic Mineral Resource Activities, done at Wellington, New Zealand, on 2 June 1988.

27 EAMREA, 1977, n. 23, p. 2.

28 United States Congress, Office of Technology Assessment, *Polar Prospects: a minerals treaty for Antarctica*, OTA–0–428, US Government Printing Office, Washington, DC, September 1989, p. 89.

Environmentalists' perspectives on the protection of Antarctica

Introduction

It is nearly thirty years since the Antarctic Treaty came into force to end military conflict and promote international scientific co-operation in the region. In the intervening years the Treaty has been augmented by several agreements and two new Conventions relating to the conservation or commercial use of the Antarctic. However, comprehensive environmental protection has not been considered until recently, perhaps because environmental threats to the Antarctic had not seemed as critical as they do now.

Public interest in protecting the Antarctic environment has exploded in recent years, for two reasons primarily. First, the negotiation of an Antarctic minerals convention has alerted the public to the dramatically increased potential for the degradation of one of the world's last wilderness regions. Second, the discovery of the ozone hole and the concern that global warming may be underway has alerted the public to the fragile status of all life on earth.

These two issues are inextricably linked in the Antarctic. If the Antarctic Treaty Parties continue to hold open the possibility of extracting fossil fuels from the Antarctic, they will signal to the world that they are not yet prepared to take the steps necessary to reduce the threat of global warming. The converse is also true; declaring Antarctica a World Park, with a ban on all mining, will be an unambiguous step towards reducing this threat. In 1990, Antarctica is indeed at the crossroads.

World Park Antarctica

Greenpeace and other members of the Antarctic and Southern Ocean Coalition (ASOC) are actively campaigning for the designation of

Antarctica as a World Park, with a comprehensive and integrated approach to regulating human activity there. An Antarctic World Park would be a conservation zone where wilderness values are paramount; where peaceful co-operation in carrying out approved activities is emphasised, and a region recognised for its role as a place to monitor global environmental pollution and atmospheric degradation. All human activities undertaken south of sixty degrees South should be judged against these values; minerals activity would be precluded because it is fundamentally incompatible with the World Park approach. Just as military activities are now prohibited under the Antarctic Treaty, ASOC proposes that mining and other environmentally destructive activities be put off limits.

Environmental importance of Antarctica

Antarctica is a continent on, around and above which live remarkable plant and wildlife communities. Birds, colonies of penguins, swarms of krill, and diverse species of fish all live in a fragile coexistence, bound by the harsh environment in which they thrive.

Antarctica is the largest wilderness on this planet, and in many ways the most fragile. It is this fragility that is one of the primary concerns should the continent ever become the focus of major development activities. The terrestrial and fresh water ecosystems of Antarctica are extremely vulnerable, because in these incredibly hard climatic conditions, growth is very slow and recovery from disturbances can take years. A footprint in a moss bed, left by a careless visitor, may remain unchanged there for a decade.

Normally an ecosystem has a wide range of levels and interrelationships. It is this complexity that gives the ecosystem stability. However, Antarctic ecosystems are simpler in some respects than those in temperate or tropical regions because they contain fewer species and food chains are consequently shorter. As a result, impacts on Antarctic ecosystems have more profound effects. One single species of krill (*Euphausia superba*) may make up about half of the zooplankton biomass, feeding seals, whales, fish, and birds. Reductions through human exploitation of any of these components of the marine ecosystem equation can cause an imbalance – an imbalance that in the Antarctic is not easily restored, by humans or by nature.

The value of Antarctica as a near-pristine wilderness is also important for conducting research on global environmental problems.

Because of the relatively few sources of local pollution as compared to other parts of the world, it is possible through Antarctic research to gauge the world-wide spread of industrial pollutants. Findings of DDT and other pesticides in Antarctic ice and air, for example, provide a clear warning which we ignore at our peril.

The importance of Antarctic research is growing as we discover that global environmental problems are becoming increasingly acute. Many climatologists are alarmed that a process of significant global climate change may be under way.[1] Chlorofluorocarbons (CFCs) are amongst the most powerful of the 'greenhouse' gases. CFCs, along with other synthetic halogenated chemicals, are also responsible for a complex series of chemical reactions which has already led to a decrease in the stratospheric ozone shield. Research undertaken in the Antarctic has been largely responsible for revealing these problems. Because of its global importance this work should have priority over other uses of Antarctica.

Perhaps most importantly, protecting the Antarctic environment in its own right has became a major public concern. Put simply, people want to know that there is one place left on earth which will not be subjected to the destruction wrought by humans in every other corner of the globe.

Environmental threats to Antarctica

There are a number of current and potential threats to the Antarctic environment stemming from the presence of human beings and their related activities in and around Antarctica. Threats to Antarctica from human activity outside the region are also of major concern, but are not included in the scope of this book.

Habitat destruction

Only about two per cent of the continent is ice-free, yet this is where Antarctic flora and fauna concentrate. Wildlife rely on these areas for breeding purposes. However, humans compete for this 'prime real estate' to build their research stations, airstrips, fuel depots and other logistical support facilities. The high price of entry into the Antarctic Treaty 'Club' – year-round maintenance of a scientific station and programme – has caused over-crowded conditions in some of the more accessible, ice-free areas. In addition, the support facilities that would be required should minerals activities begin, would result in an even greater demand for coastal land.

Improper waste disposal

Impacts to flora and fauna have already occurred as a result of current waste disposal practices. Gale force winds scatter rubbish which is not tightly secured. Furthermore, the extreme cold and lack of bacteria slow down the natural process of decay, so that human rubbish does not disappear even if out of sight. The effects of poor waste disposal procedures include lead contamination (from lead batteries and paints); wildlife entanglement and choking; the introduction of diseases (from egg and chicken products), and the introduction of alien substances to Antarctic soils.

Sewage disposal is also a problem for Antarctic bases and field parties. The most common means of disposal is flushing untreated sewage into the sea, or burial in pits.

Many stations have turned to incineration to solve their solid waste problems. However, incineration merely transfers the impact from the earth to the atmosphere, where winds can spread pollution over vast distances. The impacts of atmospheric pollution on base-line scientific studies have not been adequately considered.

Greenpeace has documented waste disposal problems at a number of bases in the Antarctic.[2] Local dumping of plastics, rubber, and batteries is common-place. McMurdo Station, the largest facility in the Antarctic, operates a landfill in a coastal area. All of these practices violate Treaty agreements, and are harmful to the environment.

Accidents

On 28 January 1989, the Argentinian vessel *Bahia Paraiso* ran aground and sank approximately two kilometres from the United States' Palmer Station. According to the US National Science Foundation, 125,000 to 150,000 gallons of fuel were spilled.[3] The ship is still lying on its side and continuing to leak fuel. Argentina has not given any indication that it intends to remedy this situation. It is estimated that approximately 67,000 gallons of diesel fuel are still onboard. The *Bahia Paraiso* is not the first ship to be lost in the treacherous waters surrounding Antarctica, nor is it likely to be the last.

Overfishing

Several important commercial finfish species in Antarctica have been heavily over-exploited. For example, there is general agreement that one population of Antarctic cod (*Notothenia rossii*) is probably now commercially extinct, being reduced to less than two and a half per cent of its pre-exploitation level. Fishing states are also harvesting krill. If effective measures are not taken to regulate this fishery, the entire Antarctic marine ecosystem could be threatened.

Mineral exploitation

The most serious threat facing Antarctica is that of mineral exploitation. The most immediate consequence would be the competition for living space between developers and wildlife. Oil development would need extensive onshore facilities in the relatively few ice-free areas of the Antarctic coastline. Indeed, it should be noted that the mere possibility of mining has already had an impact. There has been a proliferation of new bases established in the more accessible regions, with little regard for scientific needs, in a modern day version of Antarctic claim-staking.

Accidents would be inevitable under the extreme conditions posed by the Antarctic. Tankers would have to cross some of the most stormy and icebound waters of the world. More seriously, blowouts and tanker accidents would be inevitable. Chronic pollution in the normal course of drilling activities would also degrade the environment.

If minerals exploitation proceeds, not only will the environment suffer, but so will vital scientific research. First, mineral exploitation will jeopardise the near-pristine nature of the Antarctic, a feature which is vital to much of the research carried out there. Second, the spirit of co-operation will be lost as research priorities shift towards resource exploitation and results are increasingly considered to be proprietary data.

Finally, mining in the Antarctic also could have negative consequences for peace. Common sense and history tell us that disputes often occur over resources, and that mineral- or energy-hungry nations will go to great lengths to protect supplies. No minerals convention will prevent such problems from arising if the stakes are too high; the only prudent course is to eliminate the source of conflict by banning all mineral activities.

Elements of a system of conservation

ASOC believes that a World Park could be established through the implementation of a comprehensive environmental protection convention that would include the following points:

- Antarctica must be maintained as a peaceful, non-militarised, nuclear-free region;
- conservation must be the paramount consideration in all human activity;
- the wilderness qualities of the continent and the Southern Ocean must be preserved;
- all minerals activities must be banned;
- science should be the primary reason for human activity on the continent, but should be limited to that which is benign, unique to Antarctica, and/or globally significant;
- there should be a ban on the storage, discharge or incineration of toxic or radioactive waste in the Antarctic, and the use of nuclear reactors for any purpose should be forbidden;
- a management body should be established to ensure that all scientific, logistical and tourist activities are examined and regulated according to a consistent set of standards, with uniform enforcement – the burden of proof should be on those proposing or undertaking any activity to show that it will not violate the regime's principles of conservation;
- there should be observer status for public interest groups, and full opportunity for participation by public interest groups and individuals; and
- a mechanism should be found for the participation of all interested states and organisations in the negotiation of the conservation regime.

Approaches to conservation

Several approaches to improving the environmental performance of the Antarctic Treaty Parties were proposed at the October 1989 Antarctic Treaty Consultative Meeting (ATCM) in Paris. Discussion has not yet ventured beyond general concepts and principles, but these undoubtedly will be further elaborated and refined as discussions intensify. A Special Consultative Meeting will be held in Santiago in late 1990 to be followed by further discussion in 1991. A number of different approaches have been put forward for discussion. Australia

and France proposed the negotiation of a comprehensive environmental protection convention which would establish environmental
principles and a system of regulation, monitoring and enforcement
applicable to all approved activities within the Antarctic Treaty area.[4]
An institutional mechanism was proposed to ensure greater account-
ability than at present because interpretation and enforcement of the
rules would not be left to individual governments. The proposal has
received strong support from the majority of environmental organisations following the issue.

Chile has put forward a proposal which envisaged the adoption of a
new set of rules and principles, possibly in the form of a new Agreed
Measures.[5] With this approach, interpretation and enforcement would
be left up to each individual member.

The United States took a status quo approach, arguing that existing
regulations could be supplemented and modified to resolve any
environmental problems in the Antarctic.[6] The US proposal stated
that the current ATCM mechanism is the only framework for
addressing environmental issues.

New Zealand proposed to strengthen existing measures and fill the
gaps in the Antarctic Treaty System (ATS).[7] The New Zealand
proposal included a comprehensive list of environmental principles to
be upheld, but like the United States, New Zealand was clearly
wedded to the existing system of agreements in force, as well as
CRAMRA. It has since announced, however, that it will not ratify the
Convention. Non-government organisations are interested to see if
and how New Zealand's proposals on comprehensive environmental
protection change as a result.

It is worth examining the effectiveness of the current ATS in
protecting the Antarctic environment. Only then can an informed
judgement be made as to which of the above proposals is best suited to
achieve the stated objective.

Antarctic Treaty and related agreements – their effectiveness in protecting the Antarctic environment

The Antarctic Treaty
In the preamble to the Antarctic Treaty, it is stipulated that 'it is in the
interest of all mankind that Antarctica shall continue forever to be
used exclusively for peaceful purposes and shall not become the scene
or object of international discord'.

Articles II and III call for the 'freedom of scientific investigation in Antarctica and co-operation towards that end', and spell out the obligation of the Contracting Parties to promote such co-operation.

Article V prohibits any nuclear explosions in Antarctica as well as the disposal there of radioactive waste material.

Article IX calls on the parties to hold meetings for the purposes of, among other things, agreeing to new measures in furtherance of the Treaty. It lists specific new measures which might be adopted, such as the preservation and conservation of living resources. The list does not, however, include minerals resource regulation.

Compliance with the Antarctic Treaty is monitored by member states which have the right to inspect any installation. Few inspections have taken place over the last thirty years, although several nations have recently initiated such visits. In addition, apart from the annual Greenpeace visits, environmental inspections have only recently been initiated and are still relatively uncommon. No violations of the Treaty have been reported by States' inspections. Although Greenpeace inspections have not revealed any violations of the Treaty itself, numerous violations of Antarctic Treaty related rules and recommendations have been documented.

The Agreed Measures for the Conservation of Antarctic Fauna and Flora
The Agreed Measures for the Conservation of Antarctic Fauna and Flora (AMCAFF) cover the same physical area as the Antarctic Treaty, and designate the entire region as a Special Conservation Area. They prohibit the killing, wounding or capturing of any native bird or mammal (excluding whales), except with a special permit. They also state that harmful interference with the normal living conditions of any native mammal or bird should be minimised. Despite their name, the Agreed Measures do not specifically afford protection to plant life except that which is included in protected areas.

In practice, there has been no enforcement of the Agreed Measures. Environmental organisations continue to maintain that construction of the Dumont d'Urville airstrip, for example, is a breach of the Agreed Measures, and privately many government representatives agree. Publicly, however, no Treaty State has taken action to investigate these allegations despite the requirements under Articles X of both the Agreed Measures and the Antarctic Treaty.

The Convention for the Conservation of Antarctic Marine Living Resources
The Convention for the Conservation of Antarctic Marine Living
Resources (CCAMLR) is very important, both in its own right and as
an example of a potentially effective fisheries regime. It is based on an
'ecosystem approach' which means that influences throughout the
whole ecosystem are supposed to be considered when making
decisions.

The spirit of the convention has been violated, however, as
overfishing of depleted stocks continues and new fisheries develop
with no regulations in place. At its 1989 meeting, CCAMLR's
Commission failed to take the advice of its Scientific Committee;
despite the fact that most Commission members agreed that more
protection was required. The consensus decision-making process
allowed the major fishing state, the Soviet Union, to block more
stringent conservation measures.

The Convention for the Conservation of Antarctic Seals
Antarctic seals were hunted to the brink of extinction in the late
nineteenth and early twentieth centuries, but have since recovered.
The Convention for the Conservation of Antarctic Seals (CCAS)
was negotiated in 1972 in order to avoid a similar occurrence.
Unfortunately, the convention is based on the notion that the
harvesting of seals at some level is both acceptable and not damaging
to the ecosystem. On the basis of existing scientific knowledge,
however, it would be impossible to determine permissable catch
levels.

Recommendations of Antarctic Treaty Consultative Meetings
Recommendations agreed at the biennial Antarctic Treaty Consultative Party Meetings (ATCM) are non-binding on the Parties until
they are enacted into national laws. Even then, there are no guarantees
that these laws accurately reflect the intent of the Recommendations.
It is therefore difficult to enforce compliance in a uniform manner
throughout the region.

For example, waste disposal guidelines in the form of a Code of
Conduct are established through Recommendations. The 1989
ATCM approved a complete overhaul of the previous Code. While
this was welcomed, it must be noted that many countries are not
currently in compliance even with the old Code. Even if all stations
immediately were to begin complying with the new Code, there are

thirty years' accumulation of debris, including hazardous wastes, at some stations.

Convention on the Regulation of Antarctic Mineral Resource Activities
The effectiveness of the Convention on the Regulation of Antarctic Mineral Resource Activities (CRAMRA) has not been tested, nor is it likely ever to be. Several key countries have decided not to ratify the Convention, which will prevent it from coming into force. It is mentioned in this analysis because it is of central concern in the debate over whether a new environmental protection convention is needed. Without CRAMRA there are two choices facing the Treaty Parties: to leave a minerals gap in the ATS; or to negotiate a permanent mining ban in the context of the Antarctic Treaty or within a comprehensive environmental protection convention.

The Antarctic and Southern Ocean Coalition (ASOC) would prefer to see a mining ban negotiated, and indeed believe that such a ban is negotiable. The Treaty Parties have, after all, agreed to effectively demilitarise and denuclearise Antarctica. In the short term, it is certainly plausible that the Treaty Parties will expand the voluntary moratorium on mineral activities while the various approaches are being discussed. However, proponents of CRAMRA use a number of arguments to attempt to undermine support for the World Park approach.[8] The most common argument is that without CRAMRA a minerals free-for-all would commence.

Without the necessary legal framework in place in Antarctica, any attempt to extract minerals would not be feasible economically or politically. In the absence of agreed sovereignty in Antarctica, a corporation would need assured property rights before expending massive sums of money to extract minerals. This view has been expressed by many international law experts. The US State Department, for example, turned down a prospecting application in 1975 precisely because there are no procedures for granting rights in Antarctica. Sir Arthur Watts, head of the British delegation to the mineral negotiations once said: 'If minerals activities in Antarctica are ever to become a reality, sufficient political stability is a precondition and a negotiated regime is the only way to establish that stability.'[9]

It has been argued that in the absence of CRAMRA, bilateral agreements might be reached, which could threaten the delicate balance of the Antarctic Treaty. However, such agreements would have to be made before significant financial resources were invested in a project. Given the suspicion that would arise amongst the other member states, it seems unlikely that a deal would be made in advance of available hard data on mineral resources.

It is also frequently argued that CRAMRA's consensus requirement to open an area amounts to a ban on mining until all parties agree otherwise. This is simply not true. A close reading of the Convention reveals that any country wishing to veto all mining applications on the grounds that mining is fundamentally incompatible with environmental protection may not be successful.

First, Article 41(2) states that 'the Commission shall identify an area for possible exploration and development if there is a consensus of Commission members that such identification is consistent with this Convention.' A lack of consensus can, therefore, only prevent an application from proceeding if it is inconsistent with the aims of CRAMRA. It must be noted that CRAMRA was designed to regulate, not prevent, minerals activities.

Second, CRAMRA's environmental principles as set out in Article 4 are not particularly iron-clad. They are based on the notion that 'significant' environmental damage would not be allowed to occur. Since 'significant' is not defined, it is instructive to read the definition of 'damage to the Antarctic environment'. Article 1(15) defines damage as 'any impact . . . beyond that which is negligible or which has been assessed and judged to be acceptable pursuant to this Convention.' This means that some level of environmental damage is expected in the normal course of operations, and that non-negligible damage may be permitted if judged to be acceptable. Clearly, a country wishing to veto every application on the basis of environmental concern will find little help here.

It is sometimes argued that an absolute ban on mining will not work, that such things never do. One might have made the same point about the military ban in 1959, but fortunately the Treaty negotiators had the courage and vision to make the attempt. At the same time, it must be questioned whether CRAMRA will work. If mining is some fifty years off, as most observers suggest, it is completely irrelevant. It is inconceivable that a convention negotiated in 1940 to regulate mining would be used, unchanged, in the 1990s.

Finally, many argue that it is not possible to negotiate an absolute ban on mining. By the same measure, CRAMRA should have been impossible to negotiate. A mining ban should be far easier because it is neutral to the issue of claims. CRAMRA does not solve the problem of tensions over conflicting sovereignty claims, but merely delays them until its institutions are forced to cope with them.

By its very existence, CRAMRA legitimises the notion that mining can be carried out safely in the Antarctic. In addition, it would allow full-scale commercial prospecting to begin, leading to increased pressure to allow exploration to commence (assuming potentially recoverable resources are identified). For this reason, while environmentalists would prefer to see a permanent mining ban implemented, they accept, in the meantime, the simple demise of CRAMRA.

Conclusion

The Antarctic environment is not being protected by the agreements comprising the Antarctic Treaty System. Activities are presumed to be benign until proven otherwise. Compliance is expected under an 'honour' system; an approach which has been demonstrated to be inadequate. Something more is clearly needed. Greenpeace and other non-government organisations are pinning their hopes on the possibility of negotiating a new environmental protection regime for the Antarctic.

Notes

1 S. Schneider, *Global Warming: Are We Entering the Greenhouse Century?*, Sierra Club Books, USA, 1989.

2 Greenpeace International, *Expedition Report 1987–88, Greenpeace Antarctic Expedition*, Greenpeace International, UK, 1988.

3 P. A. Penhale, 'Argentine ship sinks near Palmer station', *Antarctic Journal of the US*, June 1989, pp. 5–12.

4 XV ATCM/WP/2, *A Joint Australian/French proposal in the form of a paper including a draft recommendation for ATCM XV, and Comprehensive Measures for the Protection of the Antarctic Environment and its Dependent and Associated Ecosystems*, submitted jointly by the Australian and French delegations.

5 XV ATCM/WP/7, *Working Paper submitted by the delegation of Chile*.

6 XV ATCM/WP/8, *Comprehensive measures for the protection of the Antarctic Environment and Dependent and Associated Ecosystems*, submitted by the United States of America.

7 XV ATCM/WP/4, *Working Paper on the Protection of the Antarctic Environment*, submitted by the New Zealand Delegation.

8 For counter-arguments see: Greenpeace International, *No Mining in*

Antarctica: A Minerals Convention is not needed, Greenpeace International, Netherlands, 1989.

9   G. Triggs (ed.), *The Antarctic Treaty Regime : Law, Environment and Resources*, Cambridge, 1987.

Antarctica: the legal regime

Introduction

In 1972, Christopher Beeby identified three factors which might ultimately constitute a threat to the future of the Antarctic Treaty System.[1] Eighteen years after this prognostication, it could be argued that at least one, if not two, of these factors is now threatening the continuity of the Antarctic Treaty system.[2] These are the exploitation of economic resources and the question of the exercise of jurisdiction.[3]

For as use of Antarctica has intensified, so too have the political and legal debates surrounding the control of these activities. Central to the issue of control of activities in Antarctica is the ability of states to exercise jurisdiction over those activities which they wish, or are obliged, to regulate or prohibit. It is necessary then to examine the jurisdictional provisions of the existing Antarctic Treaty framework and to examine the extent to which, if at all, the proponents of change have addressed this difficult issue.

The 1959 Antarctic Treaty

The 1959 Antarctic Treaty preserves Antarctica for peaceful purposes – it was the world's first nuclear free zone – and promotes freedom of scientific research. Its essential purpose is to preserve the territorial sovereignty status quo[4] so as to avoid conflict between claimant states, particularly in the area of overlapping claims between Argentina, Chile, and the United Kingdom, and between claimant states and those contesting their claims, notably the United States and the Soviet Union. Critical to achieving peace in the region is Article IV, which provides for the 'freezing' of territorial claims in the region without prejudice to existing rights. It provides that:

'Nothing contained in the present Treaty shall be interpreted as:

(a) a renunciation by any Contracting Party of previously asserted rights of or claims to territorial sovereignty in Antarctica;

(b) a renunciation or diminution by any Contracting Party of any basis of claim to territorial sovereignty in Antarctica which it may have whether as a result of its activities or those of its nationals in Antarctica, or otherwise;

(c) prejudicing the position of any Contracting Party as regards its recognition or non-recognition of any other State's rights of or claim or basis of claim to territorial sovereignty in Antarctica.'

In this manner the conflicting positions of claimant states, potential claimant states and non-claimant states are preserved.

The 'legal acrobatics' contained in Article IV have enabled states to interpret their Treaty obligations in a manner consonant with their juridical position on territorial sovereignty. Moreover, it is through Article IV that the Antarctic Treaty has been linked to subsequent conventions.[5] As has been pointed out elsewhere, '[i]ntriguing though these techniques are for the diplomat or lawyer, and justifiably criticised as they may be for a lack of objective legal content, these 'Articles IV' have facilitated the evolution of a framework for Antarctic regulation which is remarkable given the diametrically opposed positions of the negotiating Parties.'[6]

Related to the controversial issue of territorial sovereignty is that of jurisdiction. Not surprisingly, establishing a common basis on which parties to the 1959 Antarctic Treaty would exercise jurisdiction in Antarctica proved impossible.[7] The resulting compromise embodied in Article VIII is a 'limited jurisdictional formula' which, rather than addressing the substantive problem of the basis of jurisdiction, simply provides for immunity of observers and of exchange scientists. The compromise was put forward by the United Kingdom which sought to deal 'with those cases which in practice will be those which were more likely to cause difficulties.'[8] Article VIII also provides that in the event of a dispute regarding jurisdiction the Contracting Parties concerned 'shall immediately consult together with a view to reaching a mutually acceptable solution'; furthermore, questions relating to the exercise of jurisdiction in Antarctica are within the measures which may be considered by meetings of the Consultative Parties.

As a result of Article VIII, jurisdiction is exercised by Treaty Parties on different bases, some according to territorial jurisdiction (though with express 'immunity' granted scientists and expeditions) while others adopt the nationality principle of jurisdiction.[9]

Post-1959

Clearly this compromise solution was not severely strained while activities in Antarctica remained at the level of scientific research, to which the 1959 Antarctic Treaty is primarily addressed. However, the increasing exploitation of the resources of Antarctica clearly demonstrated the need for further regulation. For example, regulation of living and non-living resources is not mentioned in the 1959 Treaty, although Article IX does provide for measures to be adopted with respect to, *inter alia*, preservation and conservation of living resources. But, to become effective, recommendations must be adopted unanimously and even then are not binding on non-Consultative Parties to the Treaty unless such Parties specifically consent to be bound.[10] Nearly 200 Recommendations have been passed.

The general consensus on the need to address the problem of environmental protection led the Consultative Parties at their Fifteenth Consultative Meeting to adopt twenty-one Recommendations on the environment including a new Waste Disposal Code of Conduct; the extension of international marine pollution conventions to the Antarctic region; environmental monitoring as an integral part of environmental assessment procedures; management plans for specially protected areas, and the establishment of a new category of Specially Reserved Areas. All are linked with the 1959 Treaty though such matters were not expressly in contemplation at the time the Treaty was negotiated. Despite these substantive measures, however, 'the regime remains . . . a weak and shadowy structure given to horatory Recommendations rather than clear enforcement mechanisms.'[11]

A number of additional conventions have been concluded for the management and conservation of the resource potential of Antarctica – the 1972 Convention for the Conservation of Antarctic Seals; the 1980 Convention on the Conservation of Antarctic Marine Living Resources (CCAMLR); and the 1988 Convention on the Regulation of Antarctic Mineral Resource Activities (CRAMRA – known as the Minerals Convention). These have served to highlight the jurisdictional inconsistencies tolerated by the 1959 Treaty and the ever-present dispute over the existence of claims to territorial sovereignty in Antarctica. In particular, where conservation and management are at issue, the inadequacies of a fragmented system of prescriptive and enforcement jurisdiction are readily apparent.[12] None the less, these conventions are built upon the 1959 Treaty, particularly the provisions regarding jurisdiction, freezing of claims and geographic scope.[13]

The Sealing Convention

Concluded in 1972 but not in force until 1978, the objectives of the Sealing Convention are the protection, scientific study and rational use of Antarctic seals and the maintenance of a satisfactory balance within the ecological system.[14] Its only real value is as a precedent for CCAMLR, for although the harvesting of pelagic seals was clearly envisaged, such has not occurred.[15] None the less, it is interesting to note that Article II provides for jurisdiction of states over their nationals or vessels under their respective flags. This nationality-based jurisdiction, flowing from flag of registry, is not expressly reproduced in CCAMLR, though the application of the principle, at least with respect to the high seas, may be inferred. [16]

The Marine Living Resources Convention

The object of the CCAMLR is the conservation of the Antarctic marine ecosystem.[17] It thus adopts a unique ecosystem standard which requires conservation measures to take into account the inter-relationship of predator and prey and their environment. Parties to the Convention are required to assess the impacts of fishing not only a targeted species, but the effect on the ecological system as a whole. Such an assessment requires particular kinds of scientific data and information, not to mention analytical techniques. This information is then used as the basis for developing a conservation strategy with the goal of preventing irreversible changes in the ecological balance. The Convention is thus concerned with both exploitation and con-servation, and with ensuring that the former does not take precedence over the latter.

The Marine Living Resources Convention was negotiated by diplomatic conference, not solely by the Consultative Parties to the Antarctic Treaty, though throughout the central role of the Antarctic Treaty system in Antarctic affairs is emphasised. Moreover, a direct link is forged between CCAMLR and the Antarctic Treaty by Article IV of the former,[18] which binds parties to Articles IV and VI of the

Antarctic Treaty,[19] whether parties to it or not. Furthermore, Article XIII(1) obliges the Scientific Committee and the Antarctic Marine Living Resources Commission to co-operate with the Antarctic Treaty Consultative Parties 'on matters falling within the competence of the latter' to avoid conflict and ensure consistency.[20]

The Convention contains no enforcement provisions with 'strong juridical provisions', such as the ability of a State to board foreign flag vessels, inspect, arrest, and even seize the vessel, contraband, or crew. Nor are such provisions found in the Sealing Convention or the Antarctic Treaty. Article X does provide a limited role for the Commission, which shall draw the attention of any State not Party to the Convention to activities by its nationals or vessels which affect the implementation of the Convention. In addition, the Commission shall draw the attention of all Contracting Parties to any activity affecting implementation by a Contracting Party of its obligations under the Convention.[21]

Article XXI(1) employs the familiar technique of deliberate ambiguity: 'Each Contracting Party shall take the appropriate measures within its competence to ensure compliance with the provisions of the Convention . . .'. This permits Contracting Parties to rely on nationality or territoriality as a basis for the exercise of jurisdiction. However, the failure to specify particular enforcement mechanisms raises concerns regarding the adequacy and effectiveness of flag-state enforcement, particularly in the context of a Convention which adopts an ecosystem approach to the conservation and management of resources in a delicate ecosystem, damage to which affects all Contracting Parties. In this regard the provisions of Article XXII(1) are lamentably weak, simply imposing on Contracting Parties the obligation 'to exert appropriate efforts', consistent with the United Nations Charter, to the end that no one engages in any activity contrary to the objectives of the Convention.

1988 Convention on the Regulation of Antarctic Mineral Resource Activities[22]

The objective of the Minerals Convention is to ensure the effective regulation of Antarctic mineral resource activities[23] 'in the interest of the international community as a whole'. Such activities are prohibited except in accordance with the Convention,[24] which provides the means for assessing the environmental impact of such activities, determining whether they are acceptable, and, where acceptable, regulating such activities and ensuring conformity with the Convention (Article 1). Article 4 sets forth detailed principles as a guide to adjudging whether an activity is permissible, emphasising the need for information adequate to enable informed judgements to be made, and the necessity for an assessment of a proposed mineral resource activity in the context of the cumulative effect of all uses upon the Antarctic environment.

A notable feature of the Minerals Convention is its institutional framework comprising: a Commission with 'Antarctica-wide functions' and the ability to adopt measures supplementing the Convention; Regional Committees to be established for each area where exploration and development activities take place; a Scientific, Technical, and Environmental Advisory Committee to the Commission and Regulatory Committees, and Special Meetings of the Parties as a 'political counterpart' to the Advisory Committee to advise when there is an application to open an area for exploration and development.[25]

If the environmental safeguards and prohibitions on unapproved exploration and exploitation are to be effective, it is necessary to have adequate compliance or enforcement mechanisms. However, given the disputed nature of territorial sovereignty in Antarctica, the issue arises as to who will ensure such compliance, and on what jurisdictional basis. An option for a claimant state would be simply to apply its law to activities by operators carried out within 'its' territory. This solution does not commend itself to non-claimant states and risks a 'thawing' of the 'freezing' provisions of Article IV of the 1959 Antarctic Treaty, meticulously preserved in Article 9 of the Minerals Convention. Nor does nationality commend itself as a prescribed basis for the exercise of jurisdiction, as claimant states view this as damaging to their interests and, in any event, it might not necessarily cover all those working in Antarctica.[26]

Article 7 of the Minerals Convention addresses the problem of compliance with the Convention.[27] Rather than prescribing the basis upon which jurisdiction should be exercised, Article 7 leaves it to 'Each Party [to] take appropriate measures within its competence to ensure compliance with this Convention and any measures in effect pursuant to it.'[28] Depending upon how a state interprets 'competence', this formula leaves it open to states to take measures on the basis of the territorial or nationality bases for jurisdiction.[29] Article 7(9) provides that nothing contained within that Article shall affect the operation of Article VIII of the Antarctic Treaty.

The United Kingdom, the first claimant state, and indeed, the first signatory to the Antarctic Minerals Convention, to enact domestic implementing legislation, has utilised both the territorial and the nationality bases for jurisdiction. The main purpose of the Antarctic Minerals Act 1989[30] is to enable the United Kingdom to ratify the Antarctic Minerals Convention; it is designed to effect its 'domestication'.[31] The Act prohibits any activities connected with the exploration or exploitation of mineral resources in Antarctica by United Kingdom nationals, Scottish firms, and bodies incorporated under the law of any part of the United Kingdom,[32] unless engaged in authorised prospecting activities. Thus the Act enables the United Kingdom to implement its obligations under, *inter alia*, Article VII of the Minerals Convention, through the control of British persons and companies. The jurisdictional basis of this control is nationality.

However, the territorial approach is evident in the British Antarctic Territory Ordinance No. 2 of 1989. This Ordinance 'prohibits any mineral activities by anyone within the British Antarctic Territory, including its continental shelf, except prospecting activities licensed by the British Government under the Act, or by another contracting party to the Convention under a corresponding law.'[33] This 'complements the Act, and, as a matter of British Antarctic territory law, widens the scope of its basic prohibition as regards mineral activities carried on in the Territory.'[34]

Belgium, which opposes the Minerals Convention, has amended its Antarctic legislation[35] to prohibit prospecting, exploration and exploitation of minerals. The jurisdictional basis relied upon is nationality, but widely construed:

'Cette interdiction vise tant les actes accomplies par des personnes physiques de nationalité belge que les actes accomplies et les activités entreprises par des personnes morales de droit belge, soit directement, soit indirectement par l'éntremise de toute autre personne morale de droit belge ou etranger dans laquelle elles ont des intérets ou à laquelle elles sont liées contractuellement.'[36]

A final feature of the Minerals Convention that may prove a model for future environmental regulation in Antarctica is Article 8, which provides for liability of operators. Negotiations are to be commenced in late 1990 with a view to producing a Liability Protocol to elaborate the rules and procedures to be applied.

Proposals for change

A Special Meeting of the Consultative Parties is to be convened in late 1990 'to explore and discuss all proposals relating to the comprehensive protection of the Antarctic environment and its dependent and associated ecosystems'. The working papers of Australia and France,[37] Chile,[38] Sweden,[39] New Zealand,[40] and the United States[41] presented to the Fifteenth Consultative Meeting will form the basis of discussions. It is a matter of speculation whether a new convention, agreed measures, or a code of conduct, on the comprehensive protection of the environment will result from the Special Meeting in late 1990. What is clear, however, is that the form chosen will be critical for the legal effectiveness of the measures adopted.

Franco-Australian proposals

The Franco-Australian working paper calls for a 'comprehensive convention relating to the preservation of the Antarctic environment and dependent or interconnected ecosystems' which will declare Antarctica a 'wilderness reserve'. All human activities having an impact on the environment are to be regulated or prohibited. In the annex to the paper, France and Australia call for the establishment of measures 'in the appropriate legal form' consistent with the activity concerned and the general principles set out in the convention, and for 'appropriate means for prevention, intervention and monitoring'.

   In their draft recommendation, Australia and France confirm that their proposals will leave unaffected the 'freezing' provisions of Article IV of the Antarctic Treaty.[42]

New Zealand proposals

New Zealand proposes a 'properly integrated, comprehensive and internally consistent regime of environmental protection' founded on basic standards applicable to all forms of human activity in the Treaty area. The proposals attempt to build on, and strengthen, the existing Antarctic Treaty system. However, under 'Dispute Settlement Procedures', it is noted that 'the Treaty system has only minimal procedures for ensuring compliance with mandatory measures that have or will be adopted to protect the environment', the exception being the Antarctic Minerals Convention.[43] The New Zealand working paper suggests the application of binding dispute settlement procedures to actual or potential damage caused the Antarctic environment, or related ecosystems, arising from activities not regulated by the existing framework. The focus is upon a multilateral approach to inspection and enforcement with procedures for: identification of the agreed categories of disputes to be covered; convening a tribunal to hear and decide upon the dispute, and expediting hearings where a serious or potentially serious threat of harm to the environment is posed.[44]

United States' proposals

In this, the weakest of the proposals for a comprehensive environmental protection regime put forward at the Fifteenth Antarctic Treaty Consultative Meeting, the United States proposes a 'program of work' to achieve comprehensive protection of the Antarctic environment. Essentially a status quo approach, the paper recommends comprehensive measures building on the existing components of the Antarctic system. It sets forth standards and procedures to be applied to a list of activities, to which appraisal criteria are applied. In carrying out the programme of work, the United States identifies three areas of emphasis, the second of which is 'clear and enforceable obligations'.[45] This is to ensure that the legal requirements and obligations imposed on states, and all those involved in Antarctica, are clear. This entails not only setting effective standards but also ensuring that these standards are embodied in clear and enforceable obligations. The United States views this as adding significantly to the ability of Treaty Parties to ensure effective and consistent compliance with environmental protection measures.

Chilean proposals

Chile proposes the 'adoption of a set of rules of a mandatory nature that regulate all of man's activity in Antarctica' and that 'all human activity should be subordinate to a set of fundamental rules and juridical principles' which are elaborated in the working paper.[46] These 'fundamental rules and juridical principles' are to be embodied in new agreed measures, thus leaving their implementation and enforcement to the individual state.

Swedish proposals

The Swedish working paper contains an outline for a draft recommendation which purports to define the common ground of the other proposals within the context of 'the need to enhance the protection of the Antarctic environment in the framework of the Antarctic Treaty System'. This is to be achieved through, *inter alia*, the negotiation of an environmental Code of Conduct to which all human activities should be subject.[47]

The way forward

Does the future hold more of the lawyer's 'delights' of contrived ambiguities, bifocalism and restrictive interpretation? It is to be hoped not. Although in the past international law has proven a tool for avoiding conflict through the use of these drafting devices,[48] this new stage of Antarctic developments requires 'an abandonment of old legal methods based narrowly on state sovereignty and territorial jurisdiction in favour of new concepts to accommodate the conflicting demands.'[49] The previous piecemeal and functional approach to regulation of the Antarctic environment and its resources is clearly inappropriate to the regulation of an interrelated ecosystem, where activity in one area impacts upon the ecosystem as a whole.

When the fragility of that ecosystem is threatened with irreparable harm, it is in the interests of all parties interested in Antarctica – indeed, in the interests of all of human kind – to ensure that such harm is averted. Some of the provisions of the Minerals Convention may prove useful in this regard, particularly the provision for institutional inspections to ensure compliance with the Convention and the requirement that Parties to the Convention ensure that the Commission has the requisite standing to appear in national courts to pursue claims.

Moreover, the adoption of the 'framework treaty and linked protocol' approach characteristic of international environmental law is evidenced in the Minerals Convention in Article 8(7), which provides for the negotiation of a separate Protocol on Liability. Reducing the scope of the discretion exercised by Treaty Parties in implementing their obligations, and providing for standing in national courts of an institutional body representing the concerns of Antarctica as a whole (and, through its membership, the interests of all humankind in Antarctica), is a significant step towards effective regulation and enforcement of activities impacting on the Antarctic environment. The New Zealand focus on a multilateral approach to inspection and enforcement is of particular interest in this regard.

Despite these encouraging steps, the juridical foundations of Antarctica are still shaky. With the increasing involvement of non-claimant States in decision-making processes affecting Antarctica, and the remaining problem of exercising jurisdiction over non-nationals for not only criminal acts, but contractual and tortious ones, or for other matters such as employment and environmental regulation, it is increasingly clear that the deliberate jurisdictional ambiguities upon which the Antarctic Treaty system is based must be resolved. To be effective, comprehensive environmental regulation requires effective and comprehensive enforcement on a clear juridical basis. It is the task of the international lawyer, perhaps as diplomat, to articulate that basis.

Notes

1   C. Beeby, *The Antarctic Treaty*, New Zealand Institute of International Affairs, 1972, pp. 17–19, cited in F. Orrego Vicuna, *Antarctic mineral exploitation: the emerging legal framework*, Cambridge University Press, 1988, p. 94.

2   The third factor identified by Beeby is the entry into Antarctica on a large scale of a non-Treaty party which refuses to become a party or to abide by the Treaty rules. There is a danger that mining in Antarctica could occur in such circumstances where neither territorial nor nationality-based juris-diction could be exercised – for example, within the unclaimed fifteen per cent of Antarctic territory.

3   Jurisdiction may be defined as 'the capacity of the State under international law to prescribe or enforce a rule of law'. See American Law Institute, *Restatement of the Law (Second), Foreign Relations Law of the United States*, Washington, DC, 1965, p. 20.

4   For doubts as to whether Article IV can have the effect of preserving the status quo, see G. Triggs, 'The Antarctic Treaty system: some jurisdictional problems' in G. Triggs (ed.), '*The Antarctic Treaty Regime: Law, Environment and Resources*, Cambridge University Press, 1988, p. 89; and Vicuna, *Antarctic mineral exploitation*, n. 1, ch. 4.

5   See discussion of the Marine Living Resources Convention and of the Minerals Convention below.

6   Triggs, *The Antarctic Treaty Regime*, n. 4, p. 56.

7   The difficulty stems, in part, from the divergent views of the nature of jurisdiction held by common and civil lawyers. To the civil lawyer, the notion of jurisdiction is closely, and sometimes inseverably, linked to the notion of territorial sovereignty. The common lawyer, while recognising that jurisdiction is a manifestation of sovereignty, none the less accords the

latter separate status. Thus the United Kingdom suggested a system of jurisdiction based on the nationality principle, while France argued for a general power of jurisdiction exercisable over its territorial claims in Antarctica. However, any attempt to link jurisdiction with territory was fiercely resisted as a tacit acknowledgement of sovereign claims.

8   Statement by Great Britain in relation to Article VIII of the Treaty at the Plenary Committee of the Conference on Antarctica, 30 November 1959, cited in Vicuna, *Antarctic minerals exploitation*, n. 1, p. 92.

9   See the examples at nn. 33 to 35 below.

10   Hence the advantage of separately negotiated conventions open to Treaty and non-Treaty parties alike.

11   Triggs, *The Antarctic Treaty Regime*, n. 4, p. 55.

12   In Antarctica, there has always been the possibility of a dichotomy emerging between prescriptive jurisdiction and enforcement jurisdiction, given the relationship of jurisdiction to the thorny issue of sovereignty and the practical problems of enforcement arising from the very nature of the Antarctic environment. Moreover, national interests have all too often superceded general interests in conservation and management.

13   CCAMLR extends to the Antarctic Convergence (Article I); and negotiations leading to the Minerals Convention considered the problem of defining the maritime scope of its provisions in light of developments in the international law of the sea. See, for example, Article V.

14   See Preamble to the Sealing Convention.

15   F. M. Auburn, *Antarctic Law and Politics*, C. Hurst & Company, London, 1982, pp. 270–1.

16   See Vicuna, *Antarctic mineral exploitation*, note 1, pp. 112–113. Nationality is the basis for jurisdiction over observations and inspectors under CCAMLR; see Article XXIV(2)(c), reflecting Article VIII(l) of the Antarctic Treaty.

17   See M. Howard, 'The Convention on the Conservation of Antarctic Marine Living Resources: a five-year review', *International and Comparative Law Quarterly*, XXXVIII, 1989, p. 104.

18   A 'bifocal approach' to the interpretation of Article IV(2)(b) has been generally accepted: compare Triggs, *The Antarctic Treaty Regime*, note 4, at p. 94. Howard agrees with her assessment that 'the contrived result achieved by such dual interpretation creates instability'; see Howard, 'The Convention on the Conservation of Antarctic Marine Living Resources', n. 17, pp. 106–7.

19   Article IV is discussed above; Article VI establishes the geographic ambit of the Antarctic Treaty as the area south of sixty degrees South Latitude. For the definitional problems that have arisen in the context of the law of the sea; see Triggs, *The Antarctic Treaty Regime*, n. 4, pp. 88–109.

20   The Convention entered into force on 7 April 1982; in the first five years of its operation neither the Scientific Committee nor the Commission approached the ATCP on any matter: see Howard, 'The Convention on the

Conservation of Antarctic Marine Living Resources', n.17, p. 108.

21 Similar wording is found in Article 7(8) and (9) of the Minerals Convention.

22 *International Legal Materials*, XXVII, 1988, p. 859. The Minerals Convention was adopted at Wellington, New Zealand, on 2 June 1988 and open for signature from 25 November 1988 to 25 November 1989.

23 Defined in Article 1 as prospecting, exploration or development, but excluding Article III of the Antarctic Treaty, thereby preserving freedom of scientific research.

24 Article 3; note, however, that Article 37(2) permits prospecting to occur without authorisation by the institutions of the Convention, though in compliance with the Convention. The Sponsoring State must notify the Commission at least nine months in advance of any prospecting activity. Prospecting is defined in Article 1 and excludes activities such as drilling to depths exceeding twenty-five metres.

25 There are also permissive provisions for a Secretariat and Executive Secretary to be established by the Commission: Article 33.

26 A. D. Watts, 'The Antarctic Minerals Convention 1988', *International and Comparative Law Quarterly*, XXXIX, 1990, p.171.

27 There is also a special inspection regime to ensure compliance with the Convention involving not only observers under Article VII of the Antarctic Treaty but also observers who may be designated by the Commission or relevant Regulatory Committee.

28 Article 7(1), a reflection of Article XXI of CCAMLR.

29 This deliberate ambiguity is a feature of legal drafting of Antarctic provisions, permitting multiple interpretations of the same provision. See the comments on 'bifocal' interpretation, at n. 18 above. See also the comments of P. Birnie regarding the international lawyer as diplomat at n. 48 below.

30 Chapter 21, assented to 21 July 1989.

31 The entire Convention is not attached as a schedule to the Act; rather, selected provisions of the Convention dealing with liability are included which under the Act have the effect of law, whilst provisions on the definition of damage to the environment, and the uses of Antarctica to be taken into account in arriving at decisions about Antarctic mineral resource activities, are reproduced for the purposes of construing Part I of the Act.

32 This may be extended by Order in Council to include bodies incorporated under the law of the Channel Islands, Isle of Man or any colony: I. D. Hendry, 'The Antarctic Minerals Act 1989', *International and Comparative Law Quarterly*, XXXIX, 1990, p. 185.

33 Hendry, 'The Antarctic Minerals Act 1989', p. 190.

34 *Ibid.*, p. 190.

35 Loi modifiant la loi du 12 janvier 1978 relative à la protection de la faune et de la flore dans l'Antarctique, 23 October 1989, in force on publication in the *Moniteur Belge*, 20 January 1990.

36   Article 6(3). The prohibition does not extend to 'la recherche strictement scientifique' (Article 6(2)).

37   See the 'Franco-Australia Draft Working Paper on Possible Components for a Comprehensive Convention for the Preservation and Protection of Antarctica', XV ATCM/WP/3; also 'A Joint Australia/French Proposal in the Form of a Paper including a Draft Recommendation for ATCM XV'; and 'Comprehensive Measures for the Protection of the Antarctic Environment and its Dependent and Associated Ecosystems', XV ATCM/WP/2.

38   'Working paper submitted by the Delegation of Chile', XV ATCM/WP/7.

39   'Comprehensive Measures for the Protection of the Antarctic Environment and Dependent and Associated Ecosystems' (proposed outline of a draft recommendation by the Swedish Delegation), XV ATCM/WP/14.

40   'Working Paper on the Protection of the Antarctic Environment', XV ATCM/WP/4.

41   'Comprehensive Measures for the Protection of the Antarctic Environment and Dependent and Associated Ecosystems' XV ATCM/WP/8.

42   Joint Australian/French Proposal, XV ATCM/WP/2, p. 3.

43   See XV ATCM/WP/4, p. 9.

44   *Ibid.*, p. 10.

45   See XV ATCM/WP/8, pp. 4–5.

46   See XV ATCM/WP/7, pp. 6–7.

47   See XV ATCM/WP/14, pp. 1 and 4, respectively.

48   On the relevance of law to the solution of environmental problems in Antarctica, see P. Birnie, 'The role of international law in solving certain environmental conflicts' in J. E. Carroll (ed.), *International environmental diplomacy: The management and resolution of transfrontier environmental problems* , Cambridge University Press, 1988, p. 95; where she states:

'When attending a recent conference on Antarctica I was interested to hear a lawyer from a Foreign Office Legal Advisers' department state "I am a lawyer but I am, of course, also a diplomat." It was not an aspect of the legal adviser's work that I had confronted before but the relevance of that remark to the resolution of Antarctic conflicts is becoming increasingly apparent as negotiations on a minerals regime for that continent proceed. The drafting skills of lawyers are the main instrument for constructing the obligations and ambiguities by which the conflicts concerning such issues as sovereignty, the interests of the international community, the treaty parties' responsibilities and liabilities for environmental damage, are being resolved. Conflicting interests are accommodated and balanced in an Antarctic regime skilfully constructed by legal means.'

49   H. Fox, 'The relevance of Antarctica to the lawyer' in G. Triggs (ed.), *The Antarctic Treaty Regime: Law, Environment and Resources*, Cambridge University Press, 1987, p. 78.

Possible future developments

There is general agreement that the Antarctic Treaty system has been successful in maintaining international co-operation in scientific research and in keeping Antarctica free from military uses and the hazards of nuclear waste. The freezing under the Antarctic Treaty of the various claims of territorial sovereignty over the continent has been crucial to the co-operative effort. However, national interests have not been put entirely to one side. The growth in the number of Contracting Parties (in itself a welcome development) and the proliferation of scientific bases in Antarctica during the 1980s suggests that longer-term strategic, political and economic considerations are as important as scientific endeavours in the eyes of Treaty members.

The Antarctic Treaty system has also provided an important mechanism for protecting the Antarctic environment, including its flora and fauna, from the worst excesses of human activities. Moreover, the Antarctic Treaty system has shown a capacity to adapt to changing circumstances and various environmental protection measures have been introduced as the perceived need arose. This process is still continuing. A number of additional environmental protection measures were agreed at the October 1989 Fifteenth Antarctic Treaty Consultative Party Meeting in Paris. These measures included a new Waste Disposal Code of Conduct; the extension of international marine pollution conventions to the Antarctic region; establishment of environmental monitoring as a required part of environmental assessment procedures; management plans for Specially Protected Areas; creation of a new category of Specially Reserved Areas, and co-operation in the hydrographic charting of Antarctic waters with a view to reducing the possibility of shipping accidents.[1]

Despite this record, the Antarctic Treaty system has been criticised by some Governments and non-government organisations on the basis that the development of environmental protection measures have been piecemeal in nature, slow in development, dependent on a consensus being reached by Antarctic Treaty Consultative Parties, and not always fully observed in practice. In part this reflects increasing international awareness of environmental issues generally, and the importance of Antarctica in particular. The Antarctic Treaty system and the actions of member countries have come under scrutiny as never before. Nowhere has this scrutiny been more intense than in the case of potential minerals activities.

The minerals convention

The development of the Convention on the Regulation of Antarctic Mineral Resource Activities, which represents the outcome of some six years of negotiations, can be seen as either a major step forward in environmental protection, or as a significant threat to the Antarctic environment, depending on one's point of view. The introduction of the Convention appears to have been motivated by fears that unregulated exploitation of Antarctic minerals could present a threat to the Antarctic Treaty system.[2] Apart from the environmental effects of a possible unregulated scramble for mineral resources, the Treaty parties feared that disputes between claimant and non-claimant States could re-ignite the conflicts over sovereignty. As well, Treaty parties were no doubt conscious that a balance had to be struck between the direct interests of Antarctic Treaty parties and their commitment to preserving Antarctica in the interests of mankind as a whole. The Treaty Parties had previously failed to achieve agreement on introducing a permanent moratorium on minerals activities, despite debating the issue five times between 1972 and 1981.

It is generally acknowledged that the Minerals Convention represents a significant advance in environmental protection of this kind and it has a number of useful features. Article 4 of the Convention sets out the environmental standards which relate to prospecting, exploration and development activities. These standards are quite stringent and place the onus on proponents of minerals activities to show that they will not cause significant environmental damage. However, the central issue in the debate about the Convention is the underlying assumption that minerals activities may, at some time in the future, be capable of being undertaken without causing unacceptable environmental damage.

Those supporting the Convention argue that while little is currently known about possible mineral resources in Antarctica, there is no *a priori* reason to suggest that at some future time minerals activities could not take place without causing unacceptable environmental harm. In addition, they claim that it is unrealistic to expect that knowledge about Antarctic minerals potential will not steadily accrue through the normal processes of scientific investigation. Hence, they argue that it is best to institute a framework under which possible future minerals activities can be regulated and controlled. They make the point that should potentially exploitable mineral resources be found without the Minerals Convention in force, it may then prove exceedingly difficult, if not impossible, to negotiate a suitable regulatory convention with stringent environmental safeguards.

Comprehensive environmental protection

The opponents of the Minerals Convention believe that Antarctica is simply so valuable in global environment terms, and as a scientific laboratory, that any form of minerals activity poses an unacceptable risk to the fragile Antarctic environment. Recent shipping accidents in the Arctic and Antarctic regions have added some force to that argument. It is also argued that the Minerals Convention would provide an incentive for minerals activities to take place. A number of countries and non-government organisations wish to see Antarctica protected through a comprehensive environmental protection convention, or as a World Park, in which minerals activities would be banned and all forms of human activities more closely regulated to minimise any environmental impacts.

At this point one might well ask whether there is likely to be any real difference between the stringent controls on any mining that can be permitted under the Minerals Convention and an absolute ban under a comprehensive environmental protection convention? The answer is by no means clear cut. If the Minerals Convention were to enter into force, mineral resource activity could only take place in accordance with all the controls set out in the Convention. However, knowledge about the minerals potential is likely to increase much faster than under an absolute ban, where such knowledge would only accrue from normal scientific investigations which are less likely to be driven by political or purely commercial motives. If potentially economic minerals resources were to be discovered, the pressure for exploration and development could be intense under either regime. Under the Minerals Convention procedures and controls would be in place. However, if an absolute ban were in force and it was decided it should be lifted, controls would have to be developed under less than ideal circumstances. Against that, international resistance to lifting an absolute ban is likely to be greater because there would be no pre-existing presumption that mining might one day take place.

It has been argued that under the present circumstances where there are no specific rules covering minerals activities, commercial interest in Antarctica's minerals potential is unlikely to occur because of the legal and other uncertainties involved. One can not rule out the possibility that should an Antarctic mineral resource become of economic or political interest, some country or company may proceed to explore despite the legal uncertainties. Such a possibility seems extremely remote, but it serves to emphasise the need to agree on policy responses if minerals activities are not to be allowed. The success of the military and nuclear ban in Antarctica gives some cause for confidence that a minerals ban could also be negotiated and sustained.

Australia and France have proposed that a comprehensive environmental protection regime be negotiated which would ban all mining and establish Antarctica as a 'natural reserve – land of science'. Scientific and tourist activities would continue to be allowed but possibly subject to more comprehensive regulation. Australia and France envisage that a new legal instrument would be developed within the Antarctic Treaty system. Environmentalists appear to see the French–Australian proposal as being broadly akin to their World Park concept.

The issue of comprehensive environmental protection of Antarctica was discussed at the October 1989 meeting of the Antarctic Treaty Consultative Parties. The result of that meeting was an agreement to hold a Special Consultative Party Meeting on protection of the Antarctic environment in late 1990 at Santiago. That meeting will consider a range of proposals relating to comprehensive protection of the Antarctic environment and its dependent and associated ecosystems. Those proposals include working papers prepared by France and Australia, Chile, Sweden, New Zealand and the United States which are outlined in earlier chapters.

The October 1989 meeting also agreed to hold a second meeting in 1990 to consider all proposals relating to development of a Liability

Protocol, as provided for by Article 8(7) of the Minerals Convention. Under the Convention, no application for an exploration or development permit can be made until the Liability Protocol is in force for the Party concerned. The agreement to hold this second meeting suggests that those Treaty Parties supporting the Minerals Convention consider it essential to proceed with negotiations on the associated Liability Protocol as a separate issue from the comprehensive environment protection convention proposal. For them a possibly amended Minerals Convention and a strong Liability Protocol may represent an area of potential future compromise. However, countries such as France and Australia are unlikely to accept such an approach.

The opponents of the Minerals Convention have had some initial success in effectively reducing the prospects of the Minerals Convention entering into force. The Convention can only do so following ratification, acceptance, approval or accession by sixteen Antarctic Treaty Consultative Parties who participated as such in the 2 June 1988 Special Meeting at Wellington. According to the terms of the Convention, all seven claimant states, together with the USA and USSR, must be included in that sixteen. When the period for which the Convention was open for signature closed on 25 November 1989, six Consultative Parties which had participated in the Wellington meeting had not signed. They were Australia, Belgium, France, the Federal Republic of Germany, India and Italy. Unless they change their present positions, Australia and France, as claimant states, can prevent the Minerals Convention from entering into force. However, as no signatory states have yet ratified the Convention, Australia and France are unlikely to be accused of blocking the Minerals Convention for a period of two or three years – the normal time taken for ratification procedures to be completed in all countries.

Nevertheless, during 1990 there has been a flurry of activity as the two sides in the Minerals Convention debate have sought to narrow their areas of difference. There is clearly a large measure of agreement about the fundamental issues of the importance of Antarctica and the need to ensure a high level of environmental protection. However, the debate is not simply one of disagreement about means. It is also a debate about philosophy and international relationships. For example, should the world take a decision now to lock up the possible mineral resources of Antarctica without knowing the opportunity cost of doing so? Is it realistic to expect that a ban on minerals activity could be maintained indefinitely as knowledge about minerals prospectivity increases? And to what extent should the views of non-Antarctic Treaty parties, both government and non-government, be accommodated?

The outcome of the present debate is crucial for the future of humankind's attitude towards Antarctica. What then might be the future of Antarctica – exploitation or preservation?

Possible future developments

It seems certain that the growing international concern about humankind's impact on the environment generally has added a new dimension to Antarctic policy formulation. Present indications are that environmental issues are receiving increased political attention. These trends would suggest that the chances of the Minerals Convention entering into force as it currently stands are not strong. However, a suitable compromise on a comprehensive environmental protection regime may be reached. In the case of Australia, at least, the composition and declared stance of the Senate on the Convention would make it extremely difficult for any legislation seeking to ratify the Minerals Convention to be passed. What then are the possible outcomes?

It is clear that the principle of decision making by consensus is not something which the Treaty parties would wish to abandon. Signatory states to the Minerals Convention will be reluctant to try to force the hands of Australia and France by exerting additional pressure to sign and ratify the Convention. In addition, signatory states will need to take account of the apparently increasing support for the French–Australian position. We are, therefore, likely to see a relatively long period of negotiation as the Antarctic Treaty parties endeavour to find a generally acceptable compromise.

As demonstrated by the various proposals already put forward, there are a number of ways that better environmental protection of Antarctica could be achieved. The final outcome could range from a new comprehensive environmental protection convention, to new Agreed Measures under the Antarctic Treaty, or even, perhaps, a new Code of Conduct. It seems unlikely, however, that a consensus will emerge as early as the Special Consultative Party Meeting in 1990. It may be more realistic to look ahead to the Sixteenth Consultative Party Meeting scheduled for 1991 before a possible solution will be agreed. The immediate task is for those countries favouring a comprehensive environment regime to develop their proposals quickly to the stage where they can clearly show how such a regime would relate to, or possibly subsume, in whole or in part, existing environmental protection measures. It is only by developing a draft convention or similar instrument that the relative merits of the proposals can be properly assessed.

There is, of course, a danger that while the stalemate continues over the future of the Minerals Convention, the present moratorium on minerals activities could be jeopardised. The moratorium, initiated by the Ninth Consultative Meeting in 1977, and re-endorsed by the Final Act of the Fourth Special Antarctic Treaty Consultative Meeting in 1988, is conditional on the 'timely entry into force' of the Minerals Convention. It has been suggested, for example, that already one country has engaged in prospecting activity under the guise of scientific research.[3] The link between the moratorium and the Minerals Convention could readily be broken, if consensus can be reached, through a recommendation of the 1990 Special Meeting or the Sixteenth Consultative Meeting in 1991. However, it would be desirable to give the moratorium a more secure legal basis than a recommendation if it is to remain in place indefinitely.

To some extent the present debate about minerals activities has an element of unreality about it – all participants agree that any development of Antarctic minerals resources, if indeed economic deposits exist, will not occur until well into the future. The more immediate threats to the Antarctic environment are from large-scale scientific activity and the growth in Antarctic tourism. It is estimated that about 4,000 scientific and support staff are maintained in Antarctica each summer and tourists number 3,000–4,000 each year. While these numbers are small relative to the size of the continent, the local environmental impact of the necessary supply and supporting operations, and the physical infrastructure developments, is significant and can be damaging. The present regulatory measures could be enhanced by a comprehensive environment protection regime.

A further problem which must be faced ultimately is the issue of sovereignty and legal jurisdiction. What happens, for example, if a non-Treaty party decides to undertake some minerals or other environmentally harmful activities, remote as that possibility may seem? The issues of sovereignty and jurisdiction are, of course, linked as discussed in Chapter 7. While it has to date suited the Antarctic Treaty system to operate with a deal of jurisdictional ambiguity, it is desirable for such ambiguity to be resolved to ensure the future protection of the Antarctic environment.

Despite its shortcomings, however, there are no indications that the Antarctic Treaty system will not survive for the foreseeable future. The system should prove robust enough, and the commitment of the Contracting Parties strong enough, to ensure that. Earlier criticisms by non-Treaty parties that the Antarctic Treaty was an association of secretive and privileged nations have largely been met by expanding the Treaty's membership, providing non-Consultative Party member states with observer status at Consultative Party Meetings, and by improving the flow of information to the United Nations and other bodies. The thirty-nine present Antarctic Treaty Parties represent both developed and developing countries, different political persuasions, different geographic regions, and different world status – superpowers, middle and small powers. It remains open for any country to accede to the Treaty and, provided it maintains a substantial Antarctic research programme, become a Consultative Party. However, a case can be made that the criteria for obtaining Consultative Party status should be re-examined to prevent unnecessary proliferation of scientific bases. It must be open to doubt that all the present Antarctic bases are conducting high quality scientific research.

The maintenance of a strong Antarctic Treaty system would seem to be central to future environmental protection of Antarctica. It is arguable, for example, that administration by the United Nations would result in a better outcome given the divergent interests of the countries with an active involvement in the continent.[4]

Conclusion

The practical way forward for the Treaty parties in handling the minerals issue would seem to be to place the Minerals Convention to one side for the present and to agree first on extending the present moratorium on minerals activities for a specific period of years so that this element of uncertainty is removed. Such an outcome appears to be an achievable compromise while enabling negotiations on suitable and acceptable comprehensive environment protection measures to proceed. Whatever ultimately is decided about the Minerals Convention, it contains many good initiatives such as its institutions, liability protocol, and provision for standing of its Commission in National Courts, which could serve as useful models for a comprehensive environmental protection regime.

Finding a consensus answer to the question posed by the title of this book will not be easy. It is equally true, however, that it is only through achieving a consensus that appropriate protection of Antarctica can be assured.

Notes

1   See Recommendations XV-1 to XV-22 of the XV ATCM which closed on 20 October 1989.

2   See Foreign and Commonwealth Office, *Background Brief: the Antarctic Minerals Convention and its role in protecting the Antarctic Environment*, London, September 1989, p. 4.

3   Remarks by the Under Secretary of State for Foreign and Commonwealth Affairs, during debate on the third reading of the Antarctic Minerals Bill, *House of Commons Debates*, CLVII, 17 July 1989, col. 149.

4   See T. B. Millar, 'The Way Ahead' in T. B. Millar (ed.), *Australia, Britain and Antarctica*, Australian Studies Centre, London, 1986.

Further reading

(i) Recent books:

Bonner, W. N., and Walton, D. W. H. (eds.), *Key Environments – Antarctica*, Pergamon Press, 1985.

Deacon, G. E. R. *The Antarctic Circumpolar Ocean*, Cambridge University Press, 1984.

Fifield, R. *International Research in the Antarctic*, Oxford University Press, 1987.

Laws, R. M. (ed.), *Antarctic Ecology* (Two Volumes), Academic Press, 1984.

Laws, R. M. *Antarctica : The Last Frontier*, Boxtree, London, 1989.

Lovering, J. F. and Prescott, J. R. V. *Last of Lands – Antarctica*, Melbourne University Press, 1987.

May, J. *The Greenpeace Book of Antarctica: A New View of The Seventh Continent*, Dorling Kindersley, London, 1989.

Mickleburgh, E. *Beyond the Frozen Sea: Visions of Antarctica*, Bodley Head, 1987.

Parsons, A. *Antarctica : The Next Decade*, Cambridge University Press, Studies in Polar Research, 1987.

Triggs, G. (ed.), *The Antarctic Treaty Regime : Law Environment and Resources*, Cambridge University Press, Studies In Polar Research, 1987.

Walton, D. W. H. (ed.), *Antarctic Science*, Cambridge University Press, 1987.

Antarctica : Great Stories from the Frozen Continent, Reader's Digest, 1988.

(ii) Recent general articles:

Hodgson, B. 'Antarctica: A land of isolation no more', *National Geographic*, CLXXVII, April 1990, pp. 2–51.

Meikeljohn, C. 'White mischief', *LAW Magazine*, 25 September 1989, pp. 4–7.

Wilkinson, P. 'Hands off Antarctica', *Green Magazine*, January 1990, pp. 48–53; and 'Voyage to Antarctica', *Resurgence*, January/February 1990, pp. 4–7.

Appendix A

Map of Antarctica showing territorial claims

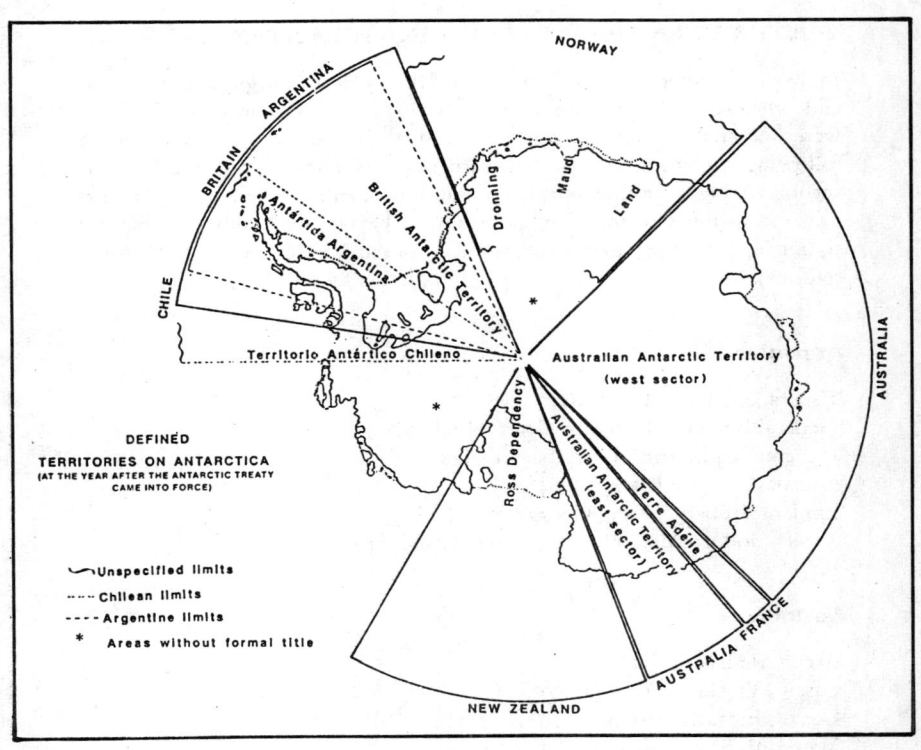

Appendix B

List and map of Antarctic bases

WINTER SCIENTIFIC STATIONS IN THE ANTARCTIC

In the 1989 winter fifty-one stations involved in meteorological observations and other scientific investigations were open in Antarctic regions. These were operated by eighteen countries and one private organisation. In the following list the numbers in square brackets refer to positions on the accompanying map; they start at the South Pole and are then in longitudinal order from the zero meridian eastbound. The symbol * indicates that the station is within the region under the aegis of the Antarctic Treaty (forty-two stations).

Argentina

'Esperanza', Hope Bay * [41]
'General Belgrano II', Caird Coast * [47]
'General San Martin', Barry Island * [28]
'Orcadas', Laurie Island * [44]
'Teniente Jubany', King George Island * [37]
'Vicecomodoro Marambio', Seymour Island * [42]

Australia

Macquarie Island, [20]
'Casey', Vincennes Bay * [18]
'Davis', Ingrid Christensen Coast * [15]
'Mawson', Mac. Robertson Land * [10]

Brazil

'Comandante Ferraz', King George Island * [39]

Britain

Bird Island, South Georgia [45]
King Edward Point, South Georgia [46]
'Faraday', Argentine Islands * [29]
'Halley', Caird Coast * [47]
'Rothera', Adelaide Island * [27]
'Signy', South Orkney Islands * [43]

Chile

'Capitán Arturo Prat', Greenwich Island * [31]
'General Bernardo O'Higgins', Cape Legoupil * [40]
'Teniente Rodolfo Marsh Martin', King George Island * [34]

China

(Peoples' Republic) 'Chang Cheng', King George Island * [32]
'Zhongshan', Princess Elizabeth Land* [12]

France

Port-aux-Français, Iles Kerguelen [11]
'Alfred-Faure', Iles Crozet [9]
'Dumont d'Urville', Terre Adélie * [19]
'Martin-de-Viviès', Ile Amsterdam, [14]

Germany (Federal Republic)

'Georg von Neumayer', Ekstrømisen * [50]

Germany (Democratic Republic)

'Georg Forster', Schirmacheroasen * [2]
'Greenpeace' station Cape Evans, Ross Island * [24]

India

'Dakshin Gangotri', Prinsesse Astrid Kyst * [4]

Japan

'Asuka', Sør-Rondane * [7]
'Syowa', Ongul * [5]

Korea (South)

'King Sejong', King George Island * [36]

New Zealand

Campbell Island [25]
'Scott Base', Ross Island * [23]

Poland

'Henryk Arctowski', King George Island * [37]

South Africa

Gough Island [49]
Marion Island [6]
'SANAE', Kronprinsesse Märtha Kyst * [51]

Soviet Union

'Bellingshausen', King George Island * [33]
'Leningradskaya', Oates Land * [21]
'Mirnyy', Queen Mary Land * [16]
'Molodezhnaya', Enderby Land * [8]
'Novolazarevskaya', Schirmacheroasen * [3]
'Progress', Princess Elizabeth Land * [13]
'Russkaya', Marie Byrd Land * [26]
'Vostok', South Geomagnetic Pole * [17]

United States

'Amundsen-Scott', South Pole * [1]
'McMurdo', Ross Island * [22]
'Palmer', Anvers Island * [30]

Uruguay

'Artigas', King George Island * [35]

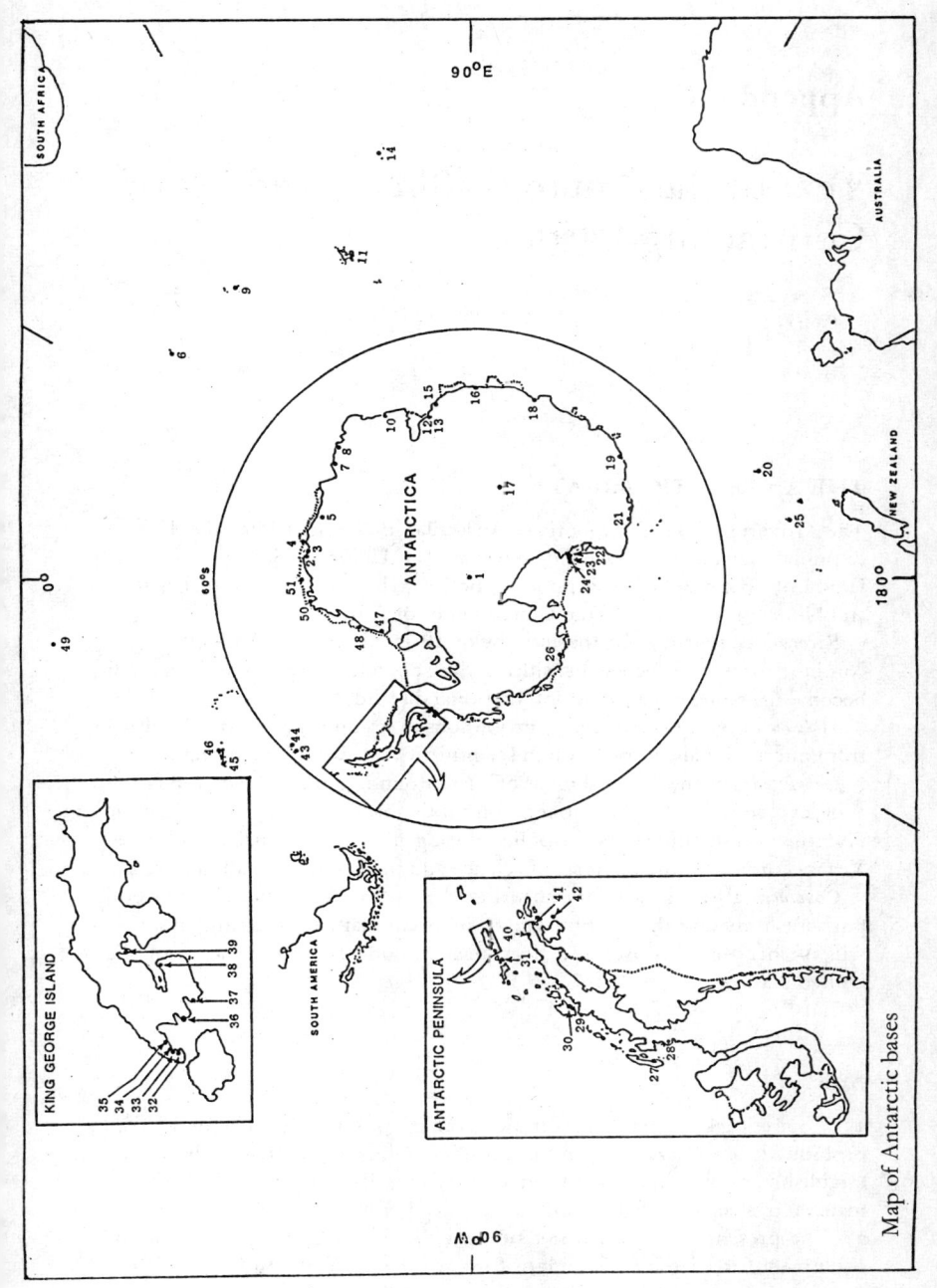

Map of Antarctic bases

Appendix C

Text of the Antarctic Treaty and list of Contracting Parties

THE ANTARCTIC TREATY

The Governments of Argentina, Australia, Belgium, Chile, the French Republic, Japan, New Zealand, Norway, the Union of South Africa, the Union of Soviet Socialist Republics, the United Kingdom of Great Britain and Northern Ireland, and the United States of America,

Recognizing that it is in the interest of all mankind that Antarctica shall continue for ever to be used exclusively for peaceful purposes and shall not become the scene or object of international discord;

Acknowledging the substantial contributions to scientific knowledge resulting from international co-operation in scientific investigation in Antarctica;

Convinced that the establishment of a firm foundation for the continuation and development of such co-operation on the basis of freedom of scientific investigation in Antarctica as applied during the International Geophysical Year accords with the interests of science and the progress of all mankind;

Convinced also that a treaty ensuring the use of Antarctica for peaceful purposes only and the continuance of international harmony in Antarctica will further the purposes and principles embodied in the Charter of the United Nations;

Have agreed as follows:

Article I

1 Antarctica shall be used for peaceful purposes only. There shall be prohibited, *inter alia*, any measure of a military nature, such as the establishment of military bases and fortifications, the carrying out of military manoeuvres, as well as the testing of any type of weapon.

2 The present Treaty shall not prevent the use of military personnel or equipment for scientific research or for any other peaceful purpose.

Article II

Freedom of scientific investigation in Antarctica and co-operation toward that end, as applied during the International Geophysical Year, shall continue, subject to the provisions of the present Treaty.

Article III

1   In order to promote international co-operation in scientific investigation in Antarctica, as provided for in Article II of the present Treaty, the Contracting Parties agree that, to the greatest extent feasible and practicable:
  (a) information regarding plans for scientific programs in Antarctica shall be exchanged to permit maximum economy of and efficiency of operations;
  (b) scientific personnel shall be exchanged in Antarctica between expeditions and stations;
  (c) scientific observations and results from Antarctica shall be exchanged and made freely available.
2   In implementing this Article, every encouragement shall be given to the establishment of co-operative working relations with those Specialized Agencies of the United Nations and other international organizations having a scientific or technical interest in Antarctica.

Article IV

1   Nothing contained in the present Treaty shall be interpreted as:
  (a) a renunciation by any Contracting Party of previously asserted rights of or claims to territorial sovereignty in Antarctica;
  (b) a renunciation or diminution by any Contracting Party of any basis of claim to territorial sovereignty in Antarctica which it may have whether as a result of its activities or those of its nationals in Antarctica, or otherwise;
  (c) prejudicing the position of any Contracting Party as regards its recognition or non-recognition of any other State's rights of or claim or basis of claim to territorial sovereignty in Antarctica.
2   No acts or activities taking place while the present Treaty is in force shall constitute a basis for asserting, supporting or denying a claim to territorial sovereignty in Antarctica or create any rights of sovereignty in Antarctica. No new claim, or enlargement of an existing claim, to territorial sovereignty in Antarctica shall be asserted while the present Treaty is in force.

Article V

1   Any nuclear explosions in Antarctica and the disposal there of radioactive waste material shall be prohibited.

2   In the event of the conclusion of international agreements concerning the use of nuclear energy, including nuclear explosions and the disposal of radioactive waste material, to which all of the Contracting Parties whose representatives are entitled to participate in the meetings provided for under Article IX are parties, the rules established under such agreements shall apply in Antarctica.

Article VI

The provisions of the present Treaty shall apply to the area south of 60° South Latitude, including all ice shelves, but nothing in the present Treaty shall prejudice or in any way affect the rights, or the exercise of the rights, of any State under international law with regard to the high seas within that area.

Article VII

1   In order to promote the objectives and ensure the observance of the provisions of the present Treaty, each Contracting Party whose representatives are entitled to participate in the meetings referred to in Article IX of the Treaty shall have the right to designate observers to carry out any inspection provided for by the present Article. Observers shall be nationals of the Contracting Parties which designate them. The names of observers shall be communicated to every other Contracting Party having the right to designate observers, and like notice shall be given of the termination of their appointment.

2   Each observer designated in accordance with the provisions of paragraph 1 of this Article shall have complete freedom of access at any time to any or all areas of Antarctica.

3   All areas of Antarctica, including all stations, installations and equipment within those areas, and all ships and aircraft at points of discharging or embarking cargoes or personnel in Antarctica, shall be open at all times to inspection by any observers designated in accordance with paragraph 1 of this Article.

4   Aerial observation may be carried out at any time over any or all areas of Antarctica by any of the Contracting Parties having the right to designate observers.

5   Each Contracting Party shall, at the time when the present Treaty enters into force for it, inform the other Contracting Parties, and thereafter shall give them notice in advance, of

(a) all expeditions to and within Antarctica, on the part of its ships or nationals, and all expeditions to Antarctica organized in or proceeding from its territory;

    (b) all stations in Antarctica occupied by its nationals; and

    (c) any military personnel or equipment intended to be introduced by it into Antarctica subject to the conditions prescribed in paragraph 2 of Article I of the present Treaty.

Article VIII

1 In order to facilitate the exercise of their functions under the present Treaty, and without prejudice to the respective positions of the Contracting Parties relating to jurisdiction over all other persons in Antarctica, observers designated under paragraph 1 of Article VII and scientific personnel exchanged under sub-paragraph 1(b) of Article III of the Treaty, and members of the staffs accompanying any such persons, shall be subject only to the jurisdiction of the Contracting Party of which they are nationals in respect of all acts or omissions occurring while they are in Antarctica for the purpose of exercising their functions.

2 Without prejudice to the provisions of paragraph 1 of this Article, and pending the adoption of measures in pursuance of sub-paragraph 1(e) of Article IX, the Contracting Parties concerned in any case of dispute with regard to the exercise of jurisdiction in Antarctica shall immediately consult together with a view to reaching a mutually acceptable solution.

Article IX

1 Representatives of the Contracting Parties named in the preamble to the present Treaty shall meet at the City of Canberra within two months after the date of entry into force of the Treaty, and thereafter at suitable intervals and places, for the purpose of exchanging information, consulting together on matters of common interest pertaining to Antarctica, and formulating and considering, and recommending to their Governments, measures in furtherance of the principles and objectives of the Treaty, including measures regarding:

    (a) use of Antarctica for peaceful purposes only;

    (b) facilitation of scientific research in Antarctica;

    (c) facilitation of international scientific co-operation in Antarctica;

    (d) facilitation of the exercise of the rights of inspection provided for in Article VII of the Treaty;

    (e) questions relating to the exercise of jurisdiction in Antarctica;

    (f) preservation and conservation of living resources in Antarctica.

2 Each Contracting Party which has become a party to the present Treaty by accession under Article XIII shall be entitled to appoint representatives to participate in the meetings referred to in paragraph 1 of the present Article, during such times as that Contracting Party demonstrates its interest in Antarctica by conducting substantial research activity there, such as the establishment of a scientific station or the despatch of a scientific expedition.

3   Reports from the observers referred to in Article VII of the present Treaty shall be transmitted to the representatives of the Contracting Parties participating in the meetings referred to in paragraph 1 of the present Article.

4   The measures referred to in paragraph 1 of this Article shall become effective when approved by all the Contracting Parties whose representatives were entitled to participate in the meetings held to consider those measures.

5   Any or all of the rights established in the present Treaty may be exercised as from the date of entry into force of the Treaty whether or not any measures facilitating the exercise of such rights have been proposed, considered or approved as provided in this Article.

Article X

Each of the Contracting Parties undertakes to exert appropriate efforts, consistent with the Charter of the United Nations, to the end that no one engages in any activity in Antarctica contrary to the principles or purposes of the present Treaty.

Article XI

1   If any dispute arises between two or more of the Contracting Parties concerning the interpretation or application of the present Treaty, those Contracting Parties shall consult among themselves with a view to having the dispute resolved by negotiation, inquiry, mediation, conciliation, arbitration, judicial settlement or other peaceful means of their own choice.

2   Any dispute of this character not so resolved shall, with the consent, in each case, of all parties to the dispute, be referred to the International Court of Justice for settlement; but failure to reach agreement on reference to the International Court shall not absolve parties to the dispute from the responsibility of continuing to seek to resolve it by any of the various peaceful means referred to in paragraph 1 of this Article.

Article XII

1   (a)   The present Treaty may be modified or amended at any time by unanimous agreement of the Contracting Parties whose representatives are entitled to participate in the meetings provided for under Article IX. Any such modification or amendment shall enter into force when the depositary Government has received notice from all such Contracting Parties that they have ratified it.

(b) Such modification or amendment shall thereafter enter into force as to any other Contracting Party when notice of ratification by it has been received by the depositary Government. Any such Contracting Party from which no notice of ratification is received within a period of two years from the date of entry into force of the modification or amendment in accordance with the provision of sub-paragraph 1(a) of this Article shall be deemed to have withdrawn from the present Treaty on the date of the expiration of such period.

2 (a) If after the expiration of thirty years from the date of entry into force of the present Treaty, any of the Contracting Parties who representatives are entitled to participate in the meetings provided for under Article IX so requests by a communication addressed to the depositary Government, a Conference of all the Contracting Parties shall be held as soon as practicable to review the operation of the Treaty.

(b) Any modification or amendment to the present Treaty which is approved at such a Conference by a majority of the Contracting Parties there represented, including a majority of those whose representatives are entitled to participate in the meetings provided for under Article IX, shall be communicated by the depositary Government to all Contracting Parties immediately after the termination of the Conference and shall enter into force in accordance with the provisions of paragraph 1 of the present Article.

(c) If any such modification or amendment has not entered into force in accordance with the provisions of sub-paragraph 1(a) of this Article within a period of two years after the date of its communication to all the Contracting Parties, any Contracting Party may at any time after the expiration of that period give notice to the depositary Government of its withdrawal from the present Treaty; and such withdrawal shall take effect two years after the receipt of the notice by the depositary Government.

Article XIII

1 The present Treaty shall be subject to ratification by the signatory States. It shall be open for accession by any State which is a Member of the United Nations, or by any other State which may be invited to accede to the Treaty with the consent of all the Contracting Parties whose representatives are entitled to participate in the meetings provided for under Article IX of the Treaty.

2 Ratification of or accession to the present Treaty shall be effected by each State in accordance with its constitutional processes.

3 Instruments of ratification and instruments of accession shall be deposited with the Government of the United States of America, hereby

designated as the depositary Government.

4 The depositary Government shall inform all signatory and acceding States of the date of each deposit of an instrument of ratification or accession, and the date of entry into force of the Treaty and of any modification or amendment thereto.

5 Upon the deposit of instruments of ratification by all the signatory States, the present Treaty shall enter into force for those States and for States which have deposited instruments of accession. Thereafter the Treaty shall enter into force for any acceding State upon the deposit of its instruments of accession.

6 The present Treaty shall be registered by the depositary Government pursuant to Article 102 of the Charter of the United Nations.

Article XIV

The present Treaty, done in the English, French, Russian and Spanish languages, each version being equally authentic, shall be deposited in the archives of the Government of the United States of America, which shall transmit duly certified copies thereof to the Governments of the signatory and acceding States.

Note

The Antarctic Treaty was signed in Washington on 1 December 1959 by twelve states, and intered into force for those states on 23 June 1961

Antarctic Treaty: Contracting Parties

(Membership in chronological order)

ᵃ United Kingdom	31 May 1960	1
ᵃ South Africa	21 June 1960	2
ᵃ Belgium	26 July 1960	3
ᵃ Japan	4 August 1960	4
ᵃ United States of America	18 August 1960	5
ᵃ Norway	24 August 1960	6
ᵃ France	16 September 1960	7
ᵃ New Zealand	1 November 1960	8
ᵃ Soviet Union	2 November 1960	9
ᵃ Poland	8 June 1961 (29 July 1977)	10
ᵃ Argentina	23 June 1961	11
ᵃ Australia	23 June 1961	12
ᵃ Chile	23 June 1961	13
Czechoslovakia	14 June 1962	14

Denmark	20 May 1965	15
Netherlands	30 March 1967	16
Romania	15 September 1971	17
[a] German Democratic Republic	19 November 1974 (5 October 1987)	18
[a] Brazil	16 May 1975 (12 September 1983)	19
Bulgaria	11 September 1978	20
[a] Germany, Federal Republic	5 February 1979 (3 March 1981)	21
[a] Uruguay	11 January 1980 (7 October 1985)	22
[b] Papua New Guinea	16 March 1981	23
[a] Italy	18 March 1981 (5 October 1987)	24
[a] Peru	10 April 1981 (9 October 1989)	25
[a] Spain	31 March 1982 (21 September 1988)	26
[a] China, Peoples' Republic	8 June 1983 (7 October 1985)	27
[a] India	19 August 1983 (12 September 1983)	28
Hungary	27 January 1984	29
[a] Sweden	24 April 1984 (21 September 1988)	30
[a] Finland	15 May 1984 (9 October 1989)	31
Cuba	16 August 1984	32
[a] Korea, Republic	28 November 1986 (9 October 1989)	33
Greece	8 January 1987	34
Korea, Democratic Republic	21 January 1987	35
Austria	25 August 1987	36
Ecuador	15 September 1987	37
Canada	4 May 1988	38
Colombia	31 January 1989	39

Original signatories: the twelve states which signed the Treaty on 1 December 1959, are *italicised*; the dates given are those of the deposition of the instruments of ratification, approval, or acceptance of the Treaty.

[a] Consultative Parties: twenty-five states, comprising the original twelve signatories and thirteen others which achieved this status after becoming actively involved in Antarctic research (with dates in parentheses).

[b] Papua New Guinea succeeded to the Treaty after achieving independence.

Appendix D

Convention on the regulation of Antarctic mineral resource activities

Preamble

The States Parties to this Convention, hereinafter referred to as the Parties,

Recalling the provisions of the Antarctic Treaty;

Convinced that the Antarctic Treaty system has proved effective in promoting international harmony in furtherance of the purposes and principles of the Charter of the United Nations, in ensuring the absence of any measures of a military nature and the protection of the Antarctic environment and in promoting freedom of scientific research in Antarctica;

Reaffirming that it is in the interest of all mankind that the Antarctic Treaty area shall continue forever to be used exclusively for peaceful purposes and shall not become the scene or object of international discord;

Noting the possibility that exploitable mineral resources may exist in Antarctica;

Bearing in mind the special legal and political status of Antarctica and the special responsibility of the Antarctic Treaty Consultative Parties to ensure that all activities in Antarctica are consistent with the purposes and principles of the Antarctic Treaty;

Bearing in mind also that a regime for Antarctic mineral resources must be consistent with Article IV of the Antarctic Treaty and in accordance therewith be without prejudice and acceptable to those States which assert rights of or claims to territorial sovereignty in Antarctica, and those States which neither recognise nor assert such rights or claims, including those States which assert a basis of claim to territorial sovereignty in Antarctica;

Noting the unique ecological, scientific and wilderness value of Antarctica and the importance of Antarctica to the global environment;

Recognising that Antarctic mineral resource activities could adversely affect the Antarctic environment or dependent or associated ecosystems;

Believing that the protection of the Antarctic environment and dependent and associated ecosystems must be a basic consideration in decisions taken on possible Antarctic mineral resource activities;

Concerned to ensure that Antarctic mineral resource activities, should they occur, are compatible with scientific investigation in Antarctica and other legitimate uses of Antarctica;

Believing that a regime governing Antarctic mineral resource activities will further strengthen the Antarctic Treaty system;

Convinced that participation in Antarctic mineral resource activities should be open to all States which have an interest in such activities and subscribe to a regime governing them and that the special situation of developing country Parties to the regime should be taken into account.

Believing that the effective regulation of Antarctic mineral resource activities is in the interest of the international community as a whole;

HAVE AGREED as follows:

CHAPTER I: GENERAL PROVISIONS

Article 1. Definitions

For the purposes of this Convention:

1 'Antarctic Treaty' means the Antarctic Treaty done at Washington on 1 December 1959.

2 'Antarctic Treaty Consultative Parties' means the Contracting Parties to the Antarctic Treaty entitled to appoint representatives to participate in the meetings referred to in Article IX of that Treaty.

3 'Antarctic Treaty area' means the area to which the provisions of the Antarctic Treaty apply in accordance with Article VI of that Treaty.

4 'Convention for the Conservation of Antarctic Seals' means the Convention done at London on 1 June 1972.

5 'Convention on the Conservation of Antarctic Marine Living Resources' means the Convention done at Canberra on 20 May 1980.

6 'Mineral resources' means all non-living natural non-renewable resources, including fossil fuels, metallic and non-metallic minerals.

7 'Antarctic mineral resource activities' means prospecting, exploration or development, but does not include scientific research activities within the meaning of Article III of the Antarctic Treaty.

8 'Prospecting' means activities, including logistic support, aimed at identifying areas of mineral resource potential for possible exploration and development, including geological, geochemical and geophysical investigations and field observations, the use of remote sensing techniques and collection of surface, seafloor and sub-ice samples. Such activities do not include dredging and excavations, except for the purpose of obtaining small-scale samples, or drilling, except shallow drilling into rock and sediment to depths not exceeding 25 metres, or such other depth as the Commission may determine for particular circumstances.

9 'Exploration' means activities, including logistic support, aimed at identifying and evaluating specific mineral resource occurrences or deposits, including exploratory drilling, dredging and other surface or subsurface

excavations required to determine the nature and size of mineral resource deposits and the feasibility of their development, but excluding pilot projects or commercial production.

10 'Development' means activities, including logistic support, which take place following exploration and are aimed at or associated with exploitation of specific mineral resource deposits, including pilot projects, processing, storage and transport activities.

11 'Operator' means:
 (a) a Party; or
 (b) an agency or instrumentality of a Party; or
 (c) a juridical person established under the law of a Party; or
 (d) a joint venture consisting exclusively of any combination of any of the foregoing,

which is undertaking Antarctic mineral resource activities and for which there is a Sponsoring State.

12 'Sponsoring State' means the Party with which an Operator has a substantial and genuine link, through being:
 (a) in the case of a Party, that Party;
 (b) in the case of an agency or instrumentality of a Party, that Party;
 (c) in the case of a juridical person other than an agency or instrumentality of a Party, the Party:
  (i) under whose law that juridical person is established and to whose law it is subject, without prejudice to any other law which might be applicable, and
  (ii) in whose territory the management of that juridical person is located, and
  (iii) to whose effective control that juridical person is subject;
 (d) in the case of a joint venture not constituting a juridical person:
  (i) where the managing member of the joint venture is a Party or an agency or instrumentality of a Party, that Party; or
  (ii) in any other case, where in relation to a Party the managing member of the joint venture satisfies the requirements of subparagraph (c) above, that Party.

13 'Managing member of the joint venture' means that member which the participating members in the joint venture have by agreement designated as having responsibility for central management of the joint venture, including the functions of organising and supervising the activities to be undertaken, and controlling the financial resources involved.

14 'Effective control' means the ability of the Sponsoring State to ensure the availability of substantial resources of the Operator for purposes connected with the implementation of this Convention, through the location of such resources in the territory of the Sponsoring State or otherwise.

15 'Damage to the Antarctic environment or dependent or associated ecosystems' means any impact on the living or non-living components of that

environment or those ecosystems, including harm to atmospheric, marine or terrestrial life, beyond that which is negligible or which has been assessed and judged to be acceptable pursuant to this Convention.

16 'Commission' means the Antarctic Mineral Resources Commission established pursuant to Article 18.

17 'Regulatory Committee' means an Antarctic Mineral Resources Regulatory Committee established pursuant to Article 29.

18 'Advisory Committee' means the Scientific, Technical and Environmental Advisory Committee established pursuant to Article 23.

19 'Special Meeting of Parties' means the Meeting referred to in Article 28.

20 'Arbitral Tribunal' means an Arbitral Tribunal constituted as provided for in the Annex, which forms an integral part of this Convention.

Article 2. Objectives and General Principles

1 This Convention is an integral part of the Antarctic Treaty system, comprising the Antarctic Treaty, the measures in effect under that Treaty, and its associated separate legal instruments, the prime purpose of which is to ensure that Antarctica shall continue forever to be used exclusively for peaceful purposes and shall not become the scene or object of international discord. The Parties provide through this Convention, the principles it establishes, the rules it prescribes, the institutions it creates and the decisions adopted pursuant to it, a means for:

- (a) assessing the possible impact on the environment of Antarctic mineral resource activities;
- (b) determining whether Antarctic mineral resource activities are acceptable;
- (c) governing the conduct of such Antarctic mineral resource activities as may be found acceptable; and
- (d) ensuring that any Antarctic mineral resource activities are undertaken in strict conformity with this Convention.

2 In implementing this Convention, the Parties shall ensure that Antarctic mineral resource activities, should they occur, take place in a manner consistent with all the components of the Antarctic Treaty system and the obligations flowing therefrom.

3 In relation to Antarctic mineral resource activities, should they occur, the Parties acknowledge the special responsibility of the Antarctic Treaty Consultative Parties for the protection of the environment and the need to:

- (a) protect the Antarctic environment and dependent and associated ecosystems;
- (b) respect Antarctica's significance for, and influence on, the global environment;
- (c) respect other legitimate uses of Antarctica;

(d) respect Antarctica's scientific value and aesthetic and wilderness qualities;

(e) ensure the safety of operations in Antarctica;

(f) promote opportunities for fair and effective participation of all Parties; and

(g) take into account the interests of the international community as a whole.

Article 3. Prohibition of Antarctic Mineral Resource Activities Outside this Convention

No Antarctic mineral resource activities shall be conducted except in accordance with this Convention and measures in effect pursuant to it and, in the case of exploration or development, with a Management Scheme approved pursuant to Article 48 or 54.

Article 4. Principles Concerning Judgments on Antarctic Mineral Resource Activities

1 Decisions about Antarctic mineral resource activities shall be based upon information adequate to enable informed judgments to be made about their possible impacts and no such activities shall take place unless this information is available for decisions relevant to those activities.

2 No Antarctic mineral resource activity shall take place until it is judged, based upon assessment of its possible impacts on the Antarctic environment and on dependent and on associated ecosystems, that the activity in question would not cause:

(a) significant adverse effects on air and water quality;

(b) significant changes in atmospheric, terrestrial or marine environments;

(c) significant changes in the distribution, abundance or productivity of populations of species of fauna or flora;

(d) further jeopardy to endangered or threatened species or populations of such species; or

(e) degradation of, or substantial risk to, areas of special biological, scientific, historic, aesthetic or wilderness significance.

3 No Antarctic mineral resource activity shall take place until it is judged, based upon assessment of its possible impacts, that the activity in question would not cause significant adverse effects on global or regional climate or weather patterns.

4 No Antarctic mineral resource activity shall take place until it is judged that:

(a) technology and procedures are available to provide for safe operations and compliance with paragraphs 2 and 3 above;

(b) there exists the capacity to monitor key environmental parameters and ecosystem components so as to identify any adverse effects of such activity and to provide for the modification of operating procedures as may be necessary in the light of the results of monitoring or increased knowledge of the Antarctic environment or dependent or associated ecosystems; and

(c) there exists the capacity to respond effectively to accidents, particularly those with potential environmental effects.

5 The judgments referred to in paragraphs 2, 3 and 4 above shall take into account the cumulative impacts of possible Antarctic mineral resource activities both by themselves and in combination with other such activities and other uses of Antarctica.

Article 5. Area of Application

1 This Convention shall, subject to paragraphs 2,3 and 4 below, apply to the Antarctic Treaty area.

2 Without prejudice to the responsibilities of Antarctic Treaty Consultative Parties under the Antarctic Treaty and measures pursuant to it, the Parties agree that this Convention shall regulate Antarctic mineral resource activities which take place on the continent of Antarctica and all Antarctic islands, including all ice shelves, south of 60° south latitude and in the seabed and subsoil of adjacent offshore areas up to the deep seabed.

3 For the purposes of this Convention 'deep seabed' means the seabed and subsoil beyond the geographic extent of the continental shelf as the term continental shelf is defined in accordance with international law.

4 Nothing in this Article shall be construed as limiting the application of other Articles of this Convention in so far as they relate to possible impacts outside the area referred to in paragraphs 1 and 2 above, including impacts on dependent or on associated ecosystems.

Article 6. Cooperation and International Participation

In the implementation of this Convention cooperation within its framework shall be promoted and encouragement given to international participation in Antarctic mineral resource activities by interested Parties which are Antarctic Treaty Consultative Parties and by other interested Parties, in particular, developing countries in either category. Such participation may be realised through the Parties themselves and their Operators.

Article 7. Compliance with this Convention

1 Each Party shall take appropriate measures within its competence to ensure compliance with this Convention and any measures in effect pursuant to it.

2 If a Party is prevented by the exercise of jurisdiction by another Party from ensuring compliance in accordance with paragraph 1 above, it shall not, to the extent that it is so prevented, bear responsibility for that failure to ensure compliance.

3 If any jurisdictional dispute related to compliance with this Convention or any measure in effect pursuant to it arises between two or more Parties, the Parties concerned shall immediately consult together with a view to reaching a mutually acceptable solution.

4 Each Party shall notify the Executive Secretary, for circulation to all other Parties, of the measures taken pursuant to paragraph 1 above.

5 Each Party shall exert appropriate efforts, consistent with the Charter of the United Nations, to the end that no one engages in any Antarctic mineral resource activities contrary to the objectives and principles of this Convention.

6 Each Party may, whenever it deems it necessary, draw the attention of the Commission to any activity which in its opinion affects the implementation of the objectives and principles of this Convention.

7 The Commission shall draw the attention of all Parties to any activity which, in the opinion of the Commission, affects the implementation of the objectives and principles of this Convention or the compliance by any Party with its obligations under this Convention and any measures in effect pursuant to it.

8 The Commission shall draw the attention of any State which is not a Party to this Convention to any activity undertaken by that State, its agencies or instrumentalities, natural or juridical persons, ships, aircraft or other means of transportation which, in the opinion of the Commission, affects the implementation of the objectives and principles of this Convention. The Commission shall inform all Parties accordingly.

9 Nothing in this Article shall affect the operation of Article 12(7) of this Convention or Article VIII of the Antarctic Treaty.

Article 8. Response Action and Liability

1 An Operator undertaking any Antarctic mineral resource activity shall take necessary and timely response action, including prevention, containment, clean up and removal measures, if the activity results in or threatens to result in damage to the Antarctic environment or dependent or associated ecosystems. The Operator, through its Sponsoring State, shall notify the Executive Secretary, for circulation to the relevant institutions of this Convention and to all parties, of action taken pursuant to this paragraph.

2 An Operator shall be strictly liable for:

    (a) damage to the Antarctic environment or dependent or associated ecosystems arising from its Antarctic mineral resource activities, including payment in the event that there has been no restoration to the status quo ante;

(b) loss of or impairment to an established use, as referred to in Article 15, or loss of or impairment to an established use of dependent or associated ecosystems, arising directly out of damage described in subparagraph (a) above;

(c) loss of or damage to property of a third party or loss of life or personal injury of a third party arising directly out of damage described in subparagraph (a) above; and

(d) reimbursement of reasonable costs by whomsoever incurred relating to necessary response action, including prevention, containment, clean up and removal measures, and action taken to restore the status quot ante where Antarctic mineral resource activities undertaken by that Operator result in or threaten to result in damage to the Antarctic environment or dependent or associated ecosystems.

3 (a) Damage of the kind referred to in paragraph 2 above which would not have occurred or continued if the Sponsoring State had carried out its obligations under this Convention with respect to its Operator shall, in accordance with international law, entail liability of that Sponsoring State. Such liability shall be limited to that portion of liability not satisfied by the Operator or otherwise.

(b) Nothing in subparagraph (a) above shall affect the application of the rules of international law applicable in the event that damage not referred to in that subparagraph would not have occurred or continued if the Sponsoring State had carried out its obligations under this Convention with respect to its Operator.

4 An Operator shall not be liable pursuant to paragraph 2 above if it proves that the damage has been caused directly by, and to the extent that it has been caused directly by:

(a) an event constituting in the circumstances of Antarctica a natural disaster an exceptional character which could not reasonably have been foreseen; or

(b) armed conflict, should it occur notwithstanding the Antarctic Treaty, or an act of terrorism directed against the activities of the Operator, against which no reasonable precautionary measures could have been effective.

5 Liability of an Operator for any loss of life, personal injury or loss of or damage to property other than that governed by this Article shall be regulated by applicable law and procedures.

6 If an Operator proves that damage has been caused totally or in part by an intentional or grossly negligent act or omission of the party seeking redress, that Operator may be relieved totally or in part from its obligation to pay compensation in respect of the damage suffered by such party.

7 (a) Further rules and procedures in respect of the provisions of liability

set out in this Article shall be elaborated through a separate Protocol which shall be adopted by consensus by the members of the Commission and shall enter into force according to the procedure provided for in Article 62 for the entry into force of this Convention.

(b) Such rules and procedures shall be designed to enhance the protection of the Antarctic environment and dependent and associated ecosystems.

(c) Such rules and procedures:

    (i)   may contain provisions for appropriate limits on liability, where such limits can be justified;

    (ii)   without prejudice to Article 57, shall prescribe means and mechanisms such as a claims tribunal or other fora by which claims against Operators pursuant to this Article may be assessed and adjudicated;

    (iii)   shall ensure that a means is provided to assist with immediate response action, and to satisfy liability under paragraph 2 above in the event, *inter alia*, that an Operator liable is financially incapable of meeting its obligation in full, that it exceeds any relevant limits of liability, that there is a defence to liability or that the loss or damage is of undetermined origin. Unless it is determined during the elaboration of the Protocol that there are other effective means of meeting these objectives, the Protocol shall establish a Fund or Funds and make provision in respect of such Fund or Funds, *inter alia*, for the following: – financing by Operators or on industry wide bases; – ensuring the permanent liquidity and mandatory supplementation thereof in the event of insufficiency; – reimbursement of costs of response action, by whomsoever incurred.

8 Nothing in paragraphs 4, 6 and 7 above or in the Protocol adopted pursuant to paragraph 7 shall affect in any way the provisions of paragraph 1 above.

9 No application for an exploration or development permit shall be made until the Protocol provided for in paragraph 7 above is in force for the Party lodging such application.

10 Each Party, pending the entry into force for it of the Protocol provided for in paragraph 7 above, shall ensure, consistently with Article 7 and in accordance with its legal system, that recourse is available in its national courts for adjudicating liability claims pursuant to paragraphs 2, 4 and 6 above against Operators which are engaged in prospecting. Such recourse shall include the adjudication of claims against any Operator it has sponsored. Each Party shall also ensure, in accordance with its legal system, that the Commission has the right to appear as a party in its national courts to pursue relevant liability claims under paragraph 2(a) above.

11 Nothing in this Article or in the Protocol provided for in paragraph 7 above shall be construed so as to:

(a) preclude the application of existing rules on liability, and the development in accordance with international law of further such rules, which may have application to either States or Operators; or

(b) affect the right of an Operator incurring liability pursuant to this Article to seek redress from another party which caused or contributed to the damage in question.

12 When compensation has been paid other than under this Convention liability under this Convention shall be offset by the amount of such payment.

Article 9. Protection of Legal Positions under the Antarctic Treaty

Nothing in this Convention and no acts or activities taking place while this Convention is in force shall:

(a) constitute a basis for asserting, supporting or denying a claim to territorial sovereignty in the Antarctic Treaty area or create any rights of sovereignty in the Antarctic Treaty area;

(b) be interpreted as a renunciation or diminution by any Party of, or as prejudicing, any right or claim or basis of claim to territorial sovereignty in Antarctica or to exercise coastal state jurisdiction under international law;

(c) be interpreted as prejudicing the position of any Party as regards its recognition or non-recognition of any such right, claim or basis of claim; or

(d) affect the provision of Article IV(2) of the Antarctic Treaty that no new claim, or enlargement of an existing claim, to territorial sovereignty in Antarctica shall be asserted while the Antarctic Treaty is in force.

Article 10. Consistency with the other Components of the Antarctic Treaty System

1 Each Party shall ensure that Antarctic mineral resource activities take place in a manner consistent with the components of the Antarctic Treaty system, including the Antarctic Treaty, the Convention for the Conservation of Antarctic Seals and the Convention on the Conservation of Antarctic Marine Living Resources and the measures in effect pursuant to those instruments.

2 The Commission shall consult and cooperate with the Antarctic Treaty Consultative Parties, the Contracting Parties to the Convention for the Conservation of Antarctic Seals, and the Commission for the Conservation of Antarctic Marine Living Resources with a view to ensuring the achievement of the objectives and principles of this Convention and avoiding any interference with the achievement of the objectives and principles of the

Antarctic Treaty, the Convention for the Conservation of Antarctic Seals or the Convention on the Conservation of Antarctic Marine Living Resources, or inconsistency between the measures in effect pursuant to those instruments and measures in effect pursuant to this Convention.

Article 11. Inspection under the Antarctic Treaty

All stations, installations and equipment, in the Antarctic Treaty area, relating to Antarctic mineral resource activities, as well as ships and aircraft supporting such activities at points of discharging or embarking cargoes or personnel at such stations and installations, shall be open at all times to inspection by observers designated under Article VII of the Antarctic Treaty for the purposes of that Treaty.

Article 12. Inspection under this Convention

1   In order to promote the objectives and principles and to ensure the observance of this Convention and measures in effect pursuant to it, all stations, installations and equipment relating to Antarctic mineral resource activities in the area in which these activities are regulated by this Convention, as well as ships and aircraft supporting such activities at points of discharging or embarking cargoes or personnel anywhere in that area shall be open at all times to inspection by:
  (a) observers designated by any member of the Commission who shall be nationals of that member; and
  (b) observers designated by the Commission or relevant Regulatory Committees.

2   Aerial inspection may be carried out at any time over the area in which Antarctic mineral resource activities are regulated by this Convention.

3   The Commission shall maintain an up-to-date list of observers designated pursuant to paragraph 1(a) and (b) above.

4   Reports from the observers shall be transmitted to the Commission and to any Regulatory Committee having competence in the area where the inspection has been carried out.

5   Observers shall avoid interference with the safe and normal operations of stations, installations and equipment visited and shall respect measures adopted by the Commission to protect confidentiality of data and information.

6   Inspections undertaken pursuant to paragraph 1(a) and (b) above shall be compatible and reinforce each other and shall not impose an undue burden on the operation of stations, installations and equipment visited.

7   In order to facilitate the exercise of their functions under this Convention, and without prejudice to the respective positions of the Parties relating

to jurisdiction over all other persons in the area in which Antarctic mineral resource activities are regulated by this Convention, observers designated under this Article shall be subject only to the jurisdiction of the Party of which they are nationals in respect of all acts or omissions occurring while they are in that area for the purpose of exercising their functions.

8 No exploration or development shall take place in an area identified pursuant to Article 41 until effective provision has been made for inspection in that area.

Article 13. Protected Areas

1 Antarctic mineral resource activities shall be prohibited in any area designated as a Specially Protected Area or a Site of Special Scientific Interest under Article IX(1) of the Antarctic Treaty. Such activities shall also be prohibited in any other area designated as a protected area in accordance with Article IX(1) of the Antarctic Treaty, except to the extent that the relevant measure provides otherwise. Pending any designation becoming effective in accordance with Article IX(4) of the Antarctic Treaty, no Antarctic mineral resource activities shall take place in any such area which would prejudice the purpose for which it was designated.

2 The Commission shall also prohibit or restrict Antarctic mineral resource activities in any area which, for historic, ecological, environmental, scientific or other reasons, it has designated as a protected area.

3 In exercising its powers under paragraph 2 above or under Article 41 the Commission shall consider whether to restrict or prohibit Antarctic mineral resource activities in any area, in addition to those referred to in paragraph 1 above, protected or set aside pursuant to provisions of other components of the Antarctic Treaty system, to ensure the purposes for which they are designated.

4 In relation to any area in which Antarctic mineral resource activities are prohibited or restricted in accordance with paragraph 1, 2 or 3 above, the Commission shall consider whether, for the purposes of Article 4(2)(e), it would be prudent, additionally, to prohibit or restrict Antarctic mineral resource activities in adjacent areas for the purpose of creating a buffer zone.

5 The Commission shall give effect to Article 10(2) in acting pursuant to paragraphs 2, 3 and 4 above.

6 The Commission shall, where appropriate, bring any decisions it takes pursuant to this Article to the attention of the Antarctic Treaty Consultative Parties, the Contracting Parties to the Convention for the Conservation of Antarctic Seals, the Commission for the Conservation of Antarctic Marine Living Resources and the Scientific Committee on Antarctic Research.

Article 14. Non-Discrimination

In the implementation of this Convention there shall be no discrimination against any Party or its Operators.

Article 15. Respect for Other Uses of Antarctica

1   Decisions about Antarctic mineral resource activities shall take into account the need to respect other established uses of Antarctica, including;
  - (a) the operation of stations and their associated installations, support facilities and equipment in Antarctica;
  - (b) scientific investigation in Antarctica and cooperation therein;
  - (c) the conservation, including rational use, of Antarctic marine living resources;
  - (d) tourism;
  - (e) the preservation of historic monuments; and
  - (f) navigation and aviation, that are consistent with the Antarctic Treaty system.

2   Antarctic mineral resource activities shall be conducted so as to respect any uses of Antarctica as referred to in paragraph 1 above.

Article 16. Availability and Confidentiality of Data and Information

Data and information obtained from Antarctic mineral resource activities shall, to the greatest extent practicable and feasible, be made freely available, provided that:
  - (a) as regards data and information of commercial value deriving from prospecting, they may be retained by the Operator in accordance with Article 37;
  - (b) as regards data and information deriving from exploration or development, the Commission shall adopt measures relating, as appropriate, to their release and to ensure the confidentiality of data and information of commercial value.

Article 17. Notifications and Provisional Exercise of Functions Executive Secretary

1   Where in this Convention there is a reference to the provision of information, a notification or a report to any institution provided for in this Convention and that institution has not been established, the information, notification or report shall be provided to the Executive Secretary who shall circulate it as required.

2   Where in this Convention a function is assigned to the Executive

Secretary and no Executive Secretary has been appointed under Article 33, that function shall be performed by the Depositary.

CHAPTER II: INSTITUTIONS

Article 18. Commission

1 There is hereby established the Antarctic Mineral Resources Commission.

2 Membership of the Commission shall be as follows:
  (a) each Party which was an Antarctic Treaty Consultative Party on the date when this Convention was opened for signature; and
  (b) each other Party during such time as it is actively engaged in substantial scientific, technical or environmental research in the area to which this Convention applies directly relevant to decisions about Antarctic mineral resource activities, particularly the assessments and judgements called for in Article 4; and
  (c) each other Party sponsoring Antarctic mineral resource exploration or development during such time as the relevant Management Scheme is in force.

3 A Party seeking to participate in the work of the Commission pursuant to subparagraph (b) or (c) above shall notify the Depositary of the basis upon which it seeks to become a member of the Commission. In the case of a Party which is not an Antarctic Treaty Consultative Party, such notification shall include a declaration of intent to abide by recommendations pursuant to Article IX(1) of the Antarctic Treaty. The Depositary shall communicate to each member of the Commission such notification and accompanying information.

4 The Commission shall consider the notification at its next meeting. In the event that a Party referred to in paragraph 2(b) above submitting a notification pursuant to paragraph 3 above is an Antarctic Treaty Consultative Party, it shall be deemed to have satisfied the requirements for Commission membership unless more than one-third of the members of the Commission object at the meeting at which such notification is considered. Any other Party submitting a notification shall be deemed to have satisfied the requirements for Commission membership if no member of the Commission objects at the meeting at which such notification is considered.

5 Each member of the Commission shall be represented by one representative who may be accompanied by alternate representatives and advisers.

6 Observer status in the Commission shall be open to any Party and to any Contracting Party to the Antarctic Treaty which is not a Party to this Convention.

Article 19. Commission Meetings

1 (a) The first meeting of the Commission, held for the purpose of taking organisational, financial and other decisions necessary for the effective functioning of this Convention and its institutions, shall be convened within six months of the entry into force of this Convention

    (b) After the Commission has held the meeting or meetings necessary to take the decisions referred to in subparagraph (a) above, the Commission shall not hold further meetings except in accordance with paragraph 2 or 3 below.

2 Meetings of the Commission shall be held within two months of:

    (a) receipt of a notification pursuant to Article 39;

    (b) a request by at least six members of the Commission; or

    (c) a request by a member of a Regulatory Committee in accordance with Article 49(1).

3 The Commission may establish a regular schedule of meetings if it determines that it is necessary for the effective functioning of this Convention.

4 Unless the Commission decides otherwise, its meetings hall be convened by the Executive Secretary.

Article 20. Commission Procedure

1 The Commission shall elect from among its members a Chairman and two Vice-Chairmen, each of whom shall be a representative of a different Party.

2 (a) Until such time as the Commission has established a regular schedule of meetings in accordance with Article 19(3), the Chairman and Vice-Chairmen shall be elected to serve for a period of two years, provided that if no meeting is held during that period they shall continue to serve until the conclusion of the first meeting held thereafter.

    (b) When a regular schedule of meetings has been established, the Chairman and Vice-Chairmen shall be elected to serve for a period of two years.

3 The Commission shall adopt its rules of procedure. Such rules may include provisions concerning the number of terms of office which the Chairman and Vice-Chairmen may serve and for the rotation of such offices.

4 The Commission may establish such subsidiary bodies as are necessary for the performance of its functions.

5 The Commission may decide to establish a permanent headquarters which shall be in New Zealand.

6 The Commission shall have legal personality and shall enjoy in the territory of each Party such legal capacity as may be necessary to perform its functions and achieve the objectives of this Convention.

7 The privileges and immunities to be enjoyed by the Commission, the Secretariat and representatives attending meetings in the territory of a Party shall be determined by agreement between the Commission and the Party concerned.

Article 21. Functions of the Commission

1 The functions of the Commission shall be :
  (a) to facilitate and promote the collection and exchange of scientific, technical and other information and research projects necessary to predict, detect and assess the possible environmental impact of Antarctic mineral resource activities, including the monitoring of key environmental parameters and ecosystem components;
  (b) to designate areas in which Antarctic mineral resource activities shall be prohibited or restricted in accordance with Article 13, and to perform the related functions assigned to it in that Article;
  (c) to adopt measures for the protection of the Antarctic environment and dependent and associated ecosystems and for the promotion of safe and effective exploration and development techniques and, as it may deem appropriate, to make available a handbook of such measures;
  (d) to determine, in accordance with Article 41, whether or not to identify an area for possible exploration and development, and to perform the related functions assigned to it in Article 42;
  (e) to adopt measures relating to prospecting applicable to all relevant Operators:
    (i) to determine for particular circumstances maximum drilling depths in accordance with Article 1(8);
    (ii) to restrict or prohibit prospecting consistently with Articles 13, 37 and 38;
  (f) to ensure the effective application of Articles 12(4), 37(7) and (8), 38(2) and 39(2), which require the submission to the Commission of information, notifications and reports;
  (g) to give advance public notice of matters upon which it is requesting the advice of the Advisory Committee;
  (h) to adopt measures relating to the availability and confidentiality of data and information, including measures pursuant to Article 16;
    (i) to elaborate the principle of non-discrimination set forth in Article 14;
  (j) to adopt measures with respect to maximum block sizes;
  (k) to perform the functions assigned to it in Article 29;
  (l) to review action by Regulatory Committees in accordance with Article 49;
  (m) to adopt measures in accordance with Articles 6 and 41(1) (d) related

to the promotion of cooperation and to participation in Antarctic mineral resource activities;

(n) to adopt general measures pursuant to Article 51(6);

(o) to take decisions on budgetary matters and adopt financial regulations in accordance with Article 35;

(p) to adopt measures regarding fees payable in connection with notifications submitted pursuant to Articles 37 and 39 and applications lodged pursuant to Articles 44 and 53, the purpose of which fees shall be to cover the administrative costs of handling such notifications and applications;

(q) to adopt measures regarding levies payable by Operators engaged in exploration and development, the principal purpose of which levies shall be to cover the costs of the institutions of this Convention;

(r) to determine in accordance with Article 35(7) the disposition of revenues, if any, accruing to the Commission which are surplus to the requirements for financing the budget pursuant to Article 35;

(s) to perform the functions assigned to it in Article 7(7) and (8);

(t) to perform the functions relating to inspection assigned to it in Article 12;

(u) to consider monitoring reports received pursuant to Article 52;

(v) to perform the functions relating to dispute settlement assigned to it in Article 59;

(w) to perform the functions relating to consultation and cooperation assigned to it in Articles 10(2) and 34;

(x) to keep under review the conduct of Antarctic mineral resource activities with a view to safeguarding the protection of the Antarctic environment in the interest of all mankind; and

(y) to perform such other functions as are provided for elsewhere in this Convention.

2   In performing its functions the Commission shall seek and take full account of the views of the Advisory Committee provided in accordance with Article 26.

3   Each measure adopted by the Commission shall specify the date on which it comes into effect.

4   The Commission shall, subject to Article 16 and measures in effect pursuant to it and paragraph 1(h) above, ensure that a publicly available record of its meetings and decisions and of information, notifications and reports submitted to it is maintained.

Article 22. Decision Making in the Commission

1   The Commission shall take decisions on matters of substance by a three-quarters majority of the members present and voting. When a question arises

as to whether a matter is one of substance or not, that matter shall be treated as one of substance unless otherwise decided by a three-quarters majority of the members present and voting.

2   Notwithstanding paragraph 1 above, consensus shall be required for the following:

(a)   the adoption of the budget and decisions on budgetary and related matters pursuant to Article 21(1)(p), (q) and (r) and Article 35(1), (2), (3), (4) and (5);

(b)   decisions taken pursuant to Article 21(1)(i);

(c)   decisions taken pursuant to Article 41(2).

3   Decisions on matters of procedure shall be taken by a simple majority of the members present and voting.

4   Nothing in this Article shall be interpreted as preventing the Commission, in taking decisions on matters of substance, from endeavouring to reach a consensus.

5   For the purposes of this Article, consensus means the absence of a formal objection. If, with respect to any decision covered by paragraph 2(c) above, the Chairman of the Commission determines that there would be such an objection he shall consult the members of the Commission. If, as a result of these consultations, the Chairman determines that an objection would remain, he shall convene those members most directly interested for the purpose of seeking to reconcile the differences and producing a generally acceptable proposal.

Article 23. Advisory Committee

1   There is hereby established the Scientific, Technical and Environmental Advisory Committee.

2   Membership of the Advisory Committee shall be open to all Parties.

3   Each member of the Advisory Committee shall be represented by one representative with suitable scientific, technical or environmental competence who may be accompanied by alternate representatives and by experts and advisers.

4   Observer status in the Advisory Committee shall be open to any Contracting Party to the Antarctic Treaty or to the Convention on the Conservation of Antarctic Marine Living Resources which is not a Party to this Convention.

Article 24. Advisory Committee Meetings

1   Unless the Commission decides otherwise, the Advisory Committee shall be convened for its first meeting within six months of the first meeting of the Commission. It shall meet thereafter as necessary to fulfil its functions

on the basis of a schedule established by the Commission.

2 Meetings of the Advisory Committee, in addition to those scheduled pursuant to paragraph 1 above, shall be convened at the request of at least six members of the Commission or pursuant to Article 40(1).

3 Unless the Commission decides otherwise, the meetings of the Advisory Committee shall be convened by the Executive Secretary.

Article 25. Advisory Committee Procedure

1 The Advisory Committee shall elect from among its members a Chairman and two Vice-Chairmen, each of whom shall be a representative of a different Party.

2 (a) Until such time as the Commission has established a schedule of meetings in accordance with Article 24(1), the Chairman and Vice-Chairmen shall be elected to serve for a period of two years, provided that if no meeting is held during that period they shall continue to serve until the conclusion of the first meeting held thereafter.

    (b) When a schedule of meetings has been established, the Chairman and Vice-Chairmen shall be elected to serve for a period of two years.

3 The Advisory Committee shall give advance public notice of its meetings and of matters to be considered at each meeting so as to permit the receipt and consideration of views on such matters from international organisations having an interest in them. For this purpose the Advisory Committee may, subject to review by the Commission, establish procedures for the transmission of relevant information to these organisations.

4 The Advisory Committee shall, by a two-thirds majority of the members present and voting, adopt its rules of procedure. Such rules may include provisions concerning the number of terms of office which the Chairman and Vice-Chairmen may serve and for the rotation of such offices. The rules of procedure and any amendments thereto shall be subject to approval by the Commission.

5 The Advisory Committee may establish such subcommittees, subject to budgetary approval, as may be necessary for the performance of its functions.

Article 26. Functions of the Advisory Committee

1 The Advisory Committee shall advise the Commission and Regulatory Committees, as required by this Convention, or as requested by them, on the scientific, technical and environmental aspects of Antarctic mineral resource activities. It shall provide a forum for consultation and cooperation concerning the collection, exchange and evaluation of information related to the scientific, technical and environmental aspects of Antarctic mineral resource activities.

2 It shall provide advice to:

(a) the Commission relating to its functions under Articles 21(1)(a) to (f), (u) and (x) and 35(7)(a) (in matters relating to scientific research) as well as on the implementation of Article 4; and

(b) Regulatory Committees with respect to:

  (i) the implementation of Article 4;

  (ii) scientific, technical and environmental aspects of Articles 43(3) and (5), 45, 47, 51, 52 and 54;

  (iii) data to be collected and reported in accordance with Articles 47 and 52; and

  (iv) the scientific, technical and environmental implications of reports and reported data provided in accordance with Articles 47 and 52.

3 It shall provide advice to the Commission and to Regulatory Committees on:

(a) criteria in respect to the judgments required under Article (4)2 and (3) for the purposes of Article 4(1);

(b) types of data and information required to carry out its functions, and how they should be collected, reported and archived;

(c) scientific research which would contribute to the base of data and information required in subparagraph (b) above;

(d) effective procedures and systems for data and information analysis, evaluation, presentation and dissemination to facilitate the judgments referred to in Article 4; and

(e) possibilities for scientific, technical and environmental cooperation amongst interested Parties which are developing countries and other Parties.

4 The Advisory Committee, in providing advice on decisions to be taken in accordance with Articles 41, 43, 45 and 54 shall, in each case, undertake a comprehensive environmental and technical assessment of the proposed actions. Such assessments shall be based on all information, and any amplifications thereof, available to the Advisory Committee, including the information provided pursuant to Articles 39(2)(e), 44(2)(b)(iii) and 53(2)(b). The assessments of the Advisory Committe shall, in each case, address the nature and scope of the decisions to be taken and shall include consideration, as appropriate, of, *inter alia:*

(a) the adequacy of existing information to enable informed judgments to be made;

(b) the nature, extent, duration and intensity of likely direct environmental impacts resulting from the proposed activity;

(c) possible indirect impacts;

(d) means and alternatives by which such direct or indirect impacts might be reduced, including environmental consequences of the alternative of not proceeding;

(e)   cumulative impacts of the proposed activity in the light of existing or planned activities;

(f)   capacity to respond effectively to accidents with potential environmental effects;

(g)   the environmental significance of unavoidable impacts; and

(h)   the probabilities of accidents and their environmental consequences.

5   In preparing its advice the Advisory Committee may seek information and advice from other scientists and experts or scientific organisations as may be required on an ad hoc basis.

6   The Advisory Committee shall, with a view to promoting international participation in Antarctic mineral resource activities as provided for in Article 6, provide advice concerning the availability to interested developing country Parties and other Parties, of the information referred to in paragraph 3 above, of training programmes related to scientific, technical and environmental matters bearing on Antarctic mineral resource activities, and of opportunities for cooperation among Parties in these programmes.

Article 27. Reporting by the Advisory Committee

The Advisory Committee shall present a report on each of its meetings to the Commission and to any relevant Regulatory Committee. The report shall cover all matters considered at the meeting and shall reflect the conclusions reached and all the views expressed by members of the Advisory Committee. The report shall be circulated by the Executive Secretary to all Parties, and to observers attending the meeting, and shall thereupon be made publicly available.

Article 28. Special Meeting of Parties

1   A Special Meeting of Parties shall, as required, be convened in accordance with Article 40(2) and shall have the functions, in relation to the identification of an area for possible exploration and development, specified in Article 40(3).

2   Membership of a Special Meeting of Parties shall be open to all Parties, each of which shall be represented by one representative who may be accompanied by alternate representatives and advisers.

3   Observer status at a Special Meeting of Parties shall be open to any Contracting Party to the Antarctic Treaty which is not a party to this Convention.

4   Each Special Meeting of Parties shall elect from among its members a Chairman and Vice-Chairmen, each of whom shall serve for the duration of that meeting. The Chairman and Vice-Chairman, shall not be representatives of the same Party.

5   The Special Meeting of Parties shall, by a two-thirds majority of the

members present and voting, adopt its rules of procedure. Until such time as this has been done the Special Meeting of Parties shall apply provisional rules of procedure drawn up by the Commission.

6 Unless the Commission decides otherwise, a Special Meeting of Parties shall be convened by the Executive Secretary and shall be held at the same venue as the meeting of the commission convened to consider the identification of an area for possible exploration and development.

Article 29. Regulatory Committees

1 All Antarctic Mineral Resources Regulatory Committee shall be established for each area identified by the Commission pursuant to Article 41.

2 Subject to paragraph 6 below, each Regulatory Committee shall consist of 10 members. Membership shall be determined by the Commission in accordance with this Article and, taking into account Article 9, shall include:

(a) the member, if any, or if there are more than one, those members of the Commission identified by reference to Article 9(b) which assert rights or claims in the Identified area;

(b) the two members of the Commission also identified by reference to Article 9(b) which assert a basis of claim in Antarctica;

(c) other members of the Commission determined in accordance with this Article so that the Regulatory Committee shall, subject to paragraph 6 below, consist, in total, of 10 members:

  (i) four members identified by reference to Article 9(b) which assert rights or claims, including the member or members, if any, referred to in subparagraph (a) above and

  (ii) six members which do not assert rights or claims as described in Article 9(b), including the two members referred to in subparagraph (b) above.

3 Upon the identification of an area in accordance with Article 41(2), the Chairman of the Commission shall, as soon as possible and in any event within 90 days, make a recommendation to the Commission concerning the membership of the Regulatory Committee. To this end the Chairman shall consult, as appropriate, with the Chairman of the Advisory committee and all members of the Commission. Such recommendation shall comply with the requirements of paragraphs 2 and 4 of this Article and shall ensure:

(a) the inclusion of members of the Commission which, whether through prospecting, scientific research or otherwise, have contributed substantial scientific, technical or environmental information relevant to the identification of the area by the commission pursuant to Article 41;

(b) adequate and equitable representation of developing country members of the Commission, having regard to the overall balance between developed and developing country members of the Commission, including at least three developing country members of the Commission;

(c) that account is taken of the value of a rotation of membership of Regulatory Committees as a further means of ensuring equitable representation of members of the Commission.

4 (a) When there are one or more members of the Regulatory Committee referred to in paragraph 2(a) above, the Chairman of the Commission shall make the recommendation in respect of paragraph 2(c)(i) above upon the nomination, if any, of such member or members which shall take into account paragraph 3 above, in particular subparagraph (b) of that paragraph.

(b) In making the recommendation in respect of paragraph 2(c)(ii) above, the Chairman of the Commission shall give full weight to the views (which shall take into account paragraph 3 above) which may be presented on behalf of those members of the commission which do not assert rights of or claims to territorial sovereignty in Antarctica and, with reference to the requirements of paragraph 3(b) above, to the views which may be presented on behalf of the developing countries among them.

5 The recommendation of the Chairman of the Commission shall be deemed to have been approved by the Commission if it does not decide otherwise at the same meeting as the recommendation is submitted. In taking any decision in accordance with this Article the Commission shall ensure that the requirements of paragraphs 2 and 3 above are complied with and that the nomination, if any, referred to in paragraph 4(a) above is given effect.

6 (a) If a member of the Commission which has sponsored prospecting in the identified area and submitted the notification pursuant to Article 39 upon which the Commission based its identification of the area pursuant to Article 41, is not a member of the Regulatory Committee by virtue of paragraphs 2 and 3 above, that member of the Commission shall be a member of the Regulatory Committee until such time as an application for an exploration permit is lodged pursuant to Article 44.

(b) If a Party lodging an application for an exploration permit pursuant to Article 44 is not a member of the Regulatory Committee by virtue of paragraphs 2 and 3 above, that Party shall be a member of the Regulatory Committee for its consideration of that application. Should such application result in approval of a Management Scheme pursuant to Article 48, the Party in question shall remain a member of the Regulatory Committee during such time as that Management Scheme is in force with the right to take part in decisions on matters affecting that Management Scheme.

7   Nothing in this Article shall be interpreted as affecting Article IV of the Antarctic Treaty.

Article 30. Regulatory Committee Procedure

1   The first meeting of each Regulatory Committee shall be convened by the Executive Secretary in accordance with Article 43(1). Each Regulatory Committee shall meet thereafter when and where necessary to fulfil its functions
2   Each member of a Regulatory Committee shall be represented by one representative who may be accompanied by alternate representatives and advisers.
3   Each Regulatory Committee shall elect from among its members a Chairman and Vice-Chairman. The Chairman and vice-Chairman shall not be representatives of the same Party.
4   Any Party may attend meetings of a Regulatory Committee as an observer.
5   Each Regulatory Committee shall adopt its rules of procedure. Such rules may include provisions concerning the period and number of terms of office which the Chairman and Vice-Chairman may serve and for the rotation of such offices.

Article 31. Functions of Regulatory Committees

1   The functions of each Regulatory Committee shall be:
   (a) to undertake the preparatory work provided for in Article 43;
   (b) to consider applications for exploration and development permits in accordance with Articles 45, 46 and 54;
   (c) to approve Management Schemes and issue exploration and development permits in accordance with Articles 47, 48 and 54;
   (d) to monitor exploration and development activities in accordance with Article 52;
   (e) to perform the functions assigned to it in Article 51;
   (f) to perform the functions relating to inspection assigned to it in Article 12;
   (g) to perform the functions relating to dispute settlement assigned to it in Article 47(r); and
   (h) to perform such other functions as are provided for elsewhere in this Convention.
2   In performing its functions each Regulatory Committee shall seek and take full account of the views of the Advisory Committee provided in accordance with Article 26.
3   Each Regulatory Committee shall, subject to Article 16 and measures in effect pursuant to it and Article 21(l)(h), ensure that a publicly available record of its decisions, and of Management Schemes in force, is maintained.

Article 32. Decision Making in Regulatory Committees

1   Decisions by a Regulatory Committee pursuant to Articles 48 and 54(5) shall be taken by a two-thirds majority of the members present and voting, which majority shall include a simple majority of those members present and voting referred to in Article 29(2)(c)(i) and also a simple majority of those members present and voting referred to in Article 29(2)(c)(ii).

2   Decisions by a Regulatory Committee pursuant to Article 43(3) and (5) shall be taken by a two-thirds majority of the members present and voting, which majority shall include at least half of those members present and voting referred to in Article 29(2)(c)(i) and also at least half of those members present and voting referred to in Article 29(2)(c)(ii).

3   Decisions on all other matters of substance shall be taken by a two-thirds majority of the members present and voting. When a question arises as to whether a matter is one of substance or not, that matter shall be treated as one of substance unless otherwise decided by a two-thirds majority of the members present and voting.

4   Decisions on matters of procedure shall be taken by a simple majority of the members present and voting.

5   Nothing in this Article shall be interpreted as preventing a Regulatory Committee, in taking decisions on matters of substance, from endeavouring to reach a consensus.

Article 33. Secretariat

1   The Commission may establish a Secretariat to serve the Commission, Regulatory Committees, the Advisory Committee, the Special Meeting of Parties and any subsidiary bodies established.

2   The Commission may appoint an Executive Secretary, who shall be the head of the Secretariat, according to such procedures and on such terms and conditions as the Commission may determine. The Executive Secretary shall serve for a four year term and may be reappointed.

3   The Commission may, with due regard to the need for efficiency and economy, authorise such staff establishment for the Secretariat as may be necessary. The Executive Secretary shall appoint, direct and supervise the staff according to such rules and procedures and on such terms and conditions as the Commission may determine.

4   The Secretariat shall perform the functions specified in this Convention and, subject to the approved budget, the tasks entrusted to it by the Commission, Regulatory Committees, the Advisory Committee and the Special Meeting of Parties.

Article 34. Cooperation with International Organisations

1   The Commission and, as appropriate, the Advisory Committee shall

cooperate with the Antarctic Treaty Consultative Parties, the Contracting Parties to the Convention for the Conservation of Antarctic Seals, the Commission for the Conservation of Antarctic Marine Living Resources, and the Scientific Committee on Antarctic Research.

2   The Commission shall cooperate with the United Nations, its relevant Specialised Agencies, and, as appropriate, any international organisation which may have competence in respect of mineral resources in areas adjacent to those covered by this Convention.

3   The Commission shall also, as appropriate, cooperate with the International Union for the Conservation of Nature and Natural Resources, and with other relevant international organisations, including non-governmental organisations, having a scientific, technical or environmental interest in Antarctica.

4   The Commission may, as appropriate, accord observer status in the Commission and in the Advisory Committee to such relevant international organisations, including non-governmental organisations, as might assist in the work of the institution in question. Observer status at a Special Meeting of Parties shall be open to such organisations as have been accorded observer status in the Commission or the Advisory Committee.

5   The Commission may enter into agreements with the organisations referred to in this Article.

Article 35. Financial Provisions

1   The Commission shall adopt a budget, on an annual or other appropriate basis, for:
  (a) its activities and the activities of Regulatory Committees, the Advisory Committee, the Special Meeting of Parties, any subsidiary bodies established and the Secretariat; and
  (b) the progressive reimbursement of any contributions paid under paragraphs 5 and 6 below whenever revenues under paragraph 4 below exceed expenditure.

2   The first draft budget shall be submitted by the Depositary at least 90 days before the first meeting of the Commission. At that meeting the Commission shall adopt its first budget and decide upon arrangements for the preparation of subsequent budgets.

3   The Commission shall adopt financial regulations.

4   Subject to paragraph 5 below, the budget shall be financed, *inter alia*, by:
  (a) fees prescribed pursuant to Articles 21(1)(p) and 43(2)(b);
  (b) levies on Operators, subject to any measures adopted by the Commission in accordance with Article 21(1)(q), pursuant to Article 47(k)(i); and
  (c) such other financial payments by Operators pursuant to Article

47(k)(ii) as may be required to be paid to the institutions of this Convention.

5   If the budget is not fully financed by revenues in accordance with paragraph 4 above, and subject to reimbursement in accordance with paragraph 1(b) above, the budget shall, to the extent of any shortfall and subject to paragraph 6 below, be financed by contributions from the members of the Commission. To this end, the Commission shall adopt as soon as possible a method of equitable sharing of contributions to the budget. The budget shall, in the meantime, to the extent of any shortfall, be financed by equal contributions from each member of the Commission.

6   In adopting the method of contributions referred to in paragraph 5 above the Commission shall consider the extent to which members of and observers at institutions of this Convention may be called upon to contribute to the costs of those institutions.

7   The Commission, in determining the disposition of revenues accruing to it, which are surplus to the requirements for financing the budget pursuant to this Article, shall:

(a)   promote scientific research in Antarctica, particularly that related to the Antarctic environment and Antarctic resources, and a wide spread of participation in such research by all Parties, in particular developing country Parties;

(b)   ensure that the interests of the members of Regulatory Committees having the most direct interest in the matter in relation to the areas in question are respected in any disposition of that surplus.

8   The finances of the Commission, Regulatory Committees, the Advisory Committee, the Special Meeting of Parties, any subsidiary bodies established and the Secretariat shall accord with the financial regulations adopted by the Commission and shall be subject to an annual audit by external auditors selected by the Commission.

9   Each member of the Commission, Regulatory Committees, the Advisory Committee, the Special Meeting of Parties and any subsidiary bodies established, as well as any observer at a meeting of any of the institutions of this Convention, shall meet its own expenses arising from attendance at meetings.

10   A member of the Commission that fails to pay its contribution for two consecutive years shall not, during the period of its continuing subsequent default, have the right to participate in the taking of decisions in any of the institutions of this Convention. If it continues to be in default for a further two consecutive years, the Commission shall decide what further action should be taken, which may include loss by that member of the right to participate in meetings of the institutions of this Convention. Such member shall resume the full enjoyment of its rights upon payment of the outstanding contributions.

11   Nothing in this Article shall be construed as prejudicing the position of

any member of a Regulatory Committee cn the outcome of consideration by the Regulatory Committee of terms and conditions in a Management Scheme pursuant to Article 47(k)(ii).

Article 36. Official and Working Languages

The official and working languages of the Commission, Regulatory Committees, the Advisory Committee, the Special Meeting of Parties and any meeting convened under Article 64 shall be English, French, Russian and Spanish.

CHAPTER III. PROSPECTING

Article 37. Prospecting

1   Prospecting shall not confer upon any Operator any right to Antarctic mineral resources.

2   Prospecting shall at all times be conducted in compliance with this Convention and with measures in effect pursuant to this Convention, but shall not require authorisation by the institutions of this Convention

3   (a)   The Sponsoring State shall ensure that its Operators undertaking prospecting maintain the necessary financial and technical means to comply with Article 8(1), and, to the extent that any such Operator fails to take response action as required in Article 8(1), shall ensure that this is undertaken.

    (b)   The Sponsoring State shall also ensure that its Operators undertaking prospecting maintain financial capacity, commensurate with the nature and level of the activity undertaken and the risks involved, to comply with Article 8(2).

4   In cases where more than one Operator is engaged in prospecting in the same general area, the Sponsoring State or States shall ensure that those Operators conduct their activities with due regard to each others' rights.

5   Where an Operator wishes to conduct prospecting in an area identified under Article 41 in which another Operator has been authorised to undertake exploration or development, the Sponsoring State shall ensure that such prospecting is carried out subject to the rights of any authorised Operator and any requirements to protect its rights specified by the relevant Regulatory Committee.

6   Each Operator shall ensure upon cessation of prospecting the removal of all installations and equipment and site rehabilitation. On the request of the Sponsoring State, the Commission may waive the obligation to remove installations and equipment.

7   The Sponsoring State shall notify the Commission at least nine months in advance of the commencement of planned prospecting. The notification shall

be accompanied by such fees as may be established by the Commission in accordance with Article 21(1)(p) and shall:

(a) identify, by reference to coordinates of latitude and longitude or identifiable geographic features, the general area in which the prospecting is to take place;

(b) broadly identify the mineral resource or resources which are to be the subject of the prospecting;

(c) describe the prospecting, including the methods to be used, and the general programme of work to be undertaken and its expected duration;

(d) provide an assessment of the possible environmental and other impacts of the prospecting, taking into account possible cumulative impacts as referred to in Article 4(5).

(e) describe the measures, including monitoring programmes, to be adopted to avoid harmful environmental consequences or undue interference with other established uses of Antarctica, and outline the measures to be put into effect in the event of any accident and contingency plans for evacuation in an emergency;

(f) provide details on the Operator and certify that it:

(i) has a substantial and genuine link with the Sponsoring State as defined in Article 1(12); and

(ii) is financially and technically qualified to carry out the proposed prospecting in accordance with this Convention; and

(g) provide such further information as may be required by measures adopted by the Commission.

8 The Sponsoring State shall subsequently provide to the Commission:

(a) notification of any changes to the information referred to in paragraph 7 above;

(b) notification of the cessation of prospecting, including removal of any installations and equipment as well as site rehabilitation; and

(c) a general annual report on the prospecting undertaken by the Operator.

9 Notifications and reports submitted pursuant to this Article shall be circulated by the Executive Secretary without delay to all Parties and observers attending Commission meetings.

10 Paragraphs 7, 8 and 9 above shall not be interpreted as requiring the disclosure of data and information of commercial value.

11 The Sponsoring State shall ensure that basic data and information of commercial value generated by prospecting are maintained in archives and may at any time release part of or all such data and information, on conditions which it shall establish, for scientific or environmental purposes.

12 The Sponsoring State shall ensure that basic data and information, other than interpretative data, generated by prospecting are made readily available when such data and information are not, or are no longer, of commercial

value and, in any event, no later than 10 years after the year the data and information were collected, unless it certifies to the Commission that the data and information continue to have commercial value. It shall review at regular intervals whether such data and information may be released and shall report the results of such reviews to the Commission.

13 The Commission may adopt measures consistent with this Article relating to the release of data and information of commercial value including requirements for certifications, the frequency of reviews and maximum time limits for extensions of the protection of such data and information.

Article 38. Consideration of Prospecting by the Commission

1 If a member of the Commission considers that a notification submitted in accordance with Article 37(7) or (8), or ongoing prospecting, causes concern as to consistency with this Convention or measures in effect pursuant thereto, that member may request the Sponsoring State to provide a clarification. If that member considers that an adequate response is not forthcoming from the Sponsoring State within a reasonable time, the member may request that the Commission be convened in accordance with Article 19(2)(b) to consider the question and take appropriate action.

2 If measures applicable to all relevant Operators are adopted by the Commission following a request made in accordance with paragraph 1 above, Sponsoring States that have submitted notifications in accordance with Article 37(7) or (8), and Sponsoring States whose Operators are conducting prospecting, shall ensure that the plans and activities of their Operators are modified to the extent necessary to conform with those measures within such time limit as the Commission may prescribe, and shall notify the Commission accordingly.

CHAPTER IV: EXPLORATION

Article 39. Requests for Identification of an Area for Possible Exploration and Development

1 Any Party may submit to the Executive Secretary a notification requesting that the Commission identify an area for possible exploration and development of a particular mineral resource or resources.

2 Any such notification shall be accompanied by such fees as may be established by the Commission in accordance with Article 21(1)(p) and shall contain:

(a) a precise delineation, including coordinates, of the area proposed for identification;

(b) specification of the resource or resources for which the area would be identified and any relevant data and information, excluding data and information of commercial value, concerning that resource or those resources, including a geological description of the proposed area;

(c) a detailed description of the physical environmental characteristics of the proposed area;

(d) a description of the likely scale of exploration and development for the resource or resources involved in the proposed area and of the methods which could be employed in such exploration and development;

(e) a detailed assessment of the environmental and other impacts of possible exploration and development for the resource or resources involved, taking into account Articles 15 and 26(4); and

(f) such other information as may be required pursuant to measures adopted by the Commission.

3 A notification under paragraph 1 above shall be referred promptly by the Executive Secretary to all Parties and shall be circulated to observers attending the meeting of the Commission to be convened pursuant to Article 19(2)(a).

Article 40. Action by the Advisory Committee and Special Meeting of Parties

1 The Advisory Committee shall meet as soon as possible after the meeting of the Commission convened pursuant to Article 19(2)(a) has commenced. The Advisory Committee shall provide advice to the Commission on the notification submitted pursuant to Article 39(1). The Commission may prescribe a time limit for the provision of such advice.

2 A Special Meeting of Parties shall meet as soon as possible after circulation of the report of the Advisory Committee and in any event not later than two months after that report has been circulated.

3 The Special Meeting of Parties shall consider whether identification of an area by the Commission in accordance with the request contained in the notification would be consistent with this Convention, and shall report thereon to the Commission as soon as possible and in any event not later than 21 days from the commencement of the meeting.

4 The report of the Special Meeting of Parties to the Commission shall reflect the conclusions reached and all the views expressed by Parties participating in the meeting.

Article 41. Action by the Commission

1 The Commission shall, as soon as possible after receipt of the report of the Special Meeting of Parties, consider whether or not it will identify an area as requested. Taking full account of the views and giving special weight to the

conclusions of the Special Meeting of Parties, and taking full account of the views and the conclusions of the Advisory Committee, the Commission shall determine whether such identification would be consistent with this Convention. For this purpose:

(a) the Commission shall ensure that an area to be identified shall be such that, taking into account all factors relevant to such identification, including the physical, geological, environmental and other characteristics of such area, it forms a coherent unit for the purposes of resource management. The Commission shall thus consider whether an area to be identified should include all or part of that which was requested in the notification and, subject to the necessary assessments having been made, adjacent areas not covered by that notification;

(b) the Commission shall consider whether there are, within an area requested or to be identified, any areas in which exploration and development are or should be prohibited or restricted in accordance with Article 13;

(c) the Commission shall specify the mineral resource or resources for which the area would be identified;

(d) the Commission shall give effect to Article 6, by elaborating opportunities for joint ventures or different forms of participation, up to a defined level, including procedures for offering such participation, in possible exploration and development, within the area, by interested Parties which are Antarctic Treaty Consultative Parties and by other interested Parties, in particular, developing countries in either category;

(e) the Commission shall prescribe any additional associated conditions necessary to ensure that an area to be identified is consistent with other provisions of this Convention and may prescribe general guidelines relating to the operational requirements for exploration and development in an area to be identified including measures establishing maximum block sizes and advice concerning related support activities; and

(f) the Commission shall give effect to the requirement in Article 59 to establish additional procedures for the settlement of disputes.

2　After it has completed its consideration in accordance with paragraph 1 above, the Commission shall identify an area for possible exploration and development if there is a consensus of Commission members that such identification is consistent with this Convention.

Article 42. Revision in the Scope of an Identified Area

1　If, after an area has been identified in accordance with Article 41, a Party requests identification of an area, all or part of which is contained within the boundaries of the area already identified but in respect of a mineral resource or resources different from any resource in respect of which the area has already been identified, the request shall be dealt with in accordance with Articles 39, 40 and 41. Should the Commission identify an area in respect of such different mineral resource or resources, it shall have regard, in addition to the requirements of Article 41(1)(a), to the desirability of specifying the boundaries of the area in such a way that it can be assigned to the Regulatory Committee with competence for the area already identified.

2   In the light of increased knowledge bearing on the effective management of the area, and after seeking the views of the Advisory Committee and the relevant Regulatory Committee, the Commission may amend the boundaries of any area it has identified. In making any such amendment the Commission shall ensure that authorised exploration and development in the area are not adversely affected. Unless there are compelling reasons for doing so, the Commission shall not amend the boundaries of an area it has identified in such a way as to involve a change in the composition of the relevant Regulatory Committee.

Article 43. Preparatory Work by Regulatory Committees

1   As soon as possible after the identification of an area pursuant to Article 41, the relevant Regulatory Committee established in accordance with Article 29 shall be convened.

2   The Regulatory Committee shall:
  (a) subject to any measures adopted by the Commission pursuant to Article 21(1)(j) relating to maximum block sizes, divide its area of competence into blocks in respect of which applications for exploration and development may be submitted and make provision for a limit in appropriate circumstances on the number of blocks to be accorded to any Party;
  (b) subject to any measures adopted by the Commission pursuant to Article 21(1)(p), establish fees to be paid with any application for an exploration or development permit lodged pursuant to Article 44 or 53;
  (c) establish periods within which applications for exploration and development may be lodged, all applications received within each such period being considered as simultaneous;
  (d) establish procedures for the handling of applications; and
  (e) determine a method of resolving competing applications which are not resolved in accordance with Article 45(4)(a), which method shall, provided that all other requirements of this Convention are satisfied and consistently with measures adopted pursuant to Article 41(1)(d), include priority for the application with the broadest participation among interested Parties which are Antarctic Treaty Consultative Parties, in particular, developing countries in either category.

3 The Regulatory Committee shall adopt guidelines which are consistent with, and which taken together with, the provisions of this Convention and measures of general applicability adopted by the Commission, as well as associated conditions and general guidelines adopted by the Commission when identifying the area, shall, by addressing the relevant items in Article 47, identify the general requirements for exploration and development in its area of competence.

4 Upon adoption of guidelines under paragraph 3 above the Executive Secretary shall, without delay, inform all members of the Commission of the decisions taken by the Regulatory Committee pursuant to paragraphs 2 and 3 above and shall make them publicly available together with relevant measures, associated conditions and general guidelines adopted by the Commission.

5 The Regulatory Committee may from time to time revise guidelines adopted under paragraph 3 above, taking into account any views of the Commission.

6 In performing its functions under paragraphs 3 and 5 above, the Regulatory Committee shall seek and take full account of the views of the Advisory Committee provided in accordance with Article 26.

Article 44. Application for an Exploration Permit

1 Following completion of the work undertaken pursuant to Article 43, any Party, on behalf of an Operator for which it is the Sponsoring State, may lodge with the Regulatory Committee an application for an exploration permit within the periods established by the Regulatory Committee pursuant to Article 43(2)(c).

2 An application shall be accompanied by the fees established by the Regulatory Committee in accordance with Article 43(2)(b) and shall contain:

(a) a detailed description of the Operator, including its managerial structure, financial composition and resources and technical expertise, and, in the case of an Operator being a joint venture, the inclusion of a detailed description of the degree to which Parties are involved in the Operator through, *inter alia*, juridical persons with which Parties have substantial and genuine links, so that each component of the joint venture can be easily attributed to a Party or Parties for the purposes of identifying the level of Antarctic mineral resource activities thereof, which description of substantial and genuine links shall include a description of equity sharing;

(b) a detailed description of the proposed exploration activities and a description in as much detail as possible of proposed development activities, including:

(i) an identification of the mineral resource or resources and the block to which the application applies;

    (ii)   a detailed explanation of how the proposed activities conform with the general requirements referred to in Article 43(3);

    (iii) a detailed assessment of the environmental and other impacts of the proposed activities, taking into account Articles 15 and 26(4); and

    (iv) a description of the capacity to respond effectively to accidents, especially those with potential environmental effects;

  (c) a certification by the Sponsoring State of the capacity of the Operator to comply with the general requirements referred to in Article 43(3).

  (d) a certification by the Sponsoring State of the technical competence and financial capacity of the Operator and that the Operator has a substantial and genuine link with it as defined in Article 1(12);

  (e) a description of the manner in which the application complies with any measures adopted by the Commission pursuant to Article 41(1)(d); and

  (f) such further information as may be required by the Regulatory Committee or in measures adopted by the Commission.

Article 45. Examination of Applications

1   The Regulatory Committee shall meet as soon as possible after an application has been lodged pursuant to Article 44, for the purpose of elaborating a Management Scheme. In performing this function it shall:

  (a) determine whether the application contains sufficient or adequate information pursuant to Article 44(2). To this end, it may at any time seek further information from the Sponsoring State consistent with Article 44(2);

  (b) consider the exploration and development activities proposed in the application, and such elaborations, revisions or adaptations as necessary:

    (i)   to ensure their consistency with this Convention as well as measures in effect pursuant thereto and the general requirements referred to in Article 43(3); and

    (ii) to prescribe the specific terms and conditions of a Management Scheme in accordance with Article 47.

2   At any time during the process of consideration described above, the Regulatory Committee may decline the application if it considers that the activities proposed therein cannot be elaborated, revised or adapted to ensure consistency with this Convention as well as measures in effect pursuant thereto and the general requirements referred to in Article 43(3).

3   In performing its functions under this Article, the Regulatory Committee shall seek and take full account of the views of the Advisory Committee. To that end the Regulatory Committee shall refer to the Advisory Committee all parts of the application which are necessary for it to provide advice pursuant

to Article 26, together with any other relevant information.

4 If two or more applications meeting the requirements of Article 44(2) are lodged in respect of the same block:

(a) the competing applicants shall be invited by the Regulatory Committee to resolve the competition amongst themselves, by means of their own choice within a prescribed period;

(b) if the competition is not resolved pursuant to subparagraph (a) above it shall be resolved by the Regulatory Committee in accordance with the method determined by it pursuant to Article 43(2)(e).

Article 46. Management Scheme

In performing its functions under Article 45, including the preparation of a Management Scheme, and under Article 54, the Regulatory Committee shall have recourse to the Sponsoring State and the member or members, if any, referred to in Article 29(2)(a) and, as may be required, one or two additional members of the Regulatory Committee.

Article 47. Scope of the Management Scheme

The Management Scheme shall prescribe the specific terms and conditions for exploration and development of the mineral resource or resources concerned within the relevant block. Such terms and conditions shall be consistent with the general requirements referred to in Article 43(3), and shall cover, *inter alia*:

(a) duration of exploration and development permits;

(b) measures and procedures for the protection of the Antarctic environment and dependent and associated ecosystems, including methods, activities and undertakings by the Operator to minimise environmental risks and damage;

(c) provision for necessary and timely response action, including prevention, containment and clean up and removal measures, for restoration to the status quo ante, and for contingency plans, resources and equipment to enable such action to be taken;

(d) procedures for the implementation of different stages of exploration and development;

(e) performance requirements;

(f) technical and safety specifications, including standards and procedures to ensure safe operations;

(g) monitoring and inspection;

(h) liability;

(i) procedures for the development of mineral deposits which extend outside the area covered by a permit;

(j) resource conservation requirements;

(k) financial obligations of the Operator including:
   (i)   levies in accordance with measures adopted pursuant to Article 21(1)(q);
   (ii) payments in the nature of and similar to taxes, royalties or payments in kind;
(l) financial guarantees and insurance;
(m) assignment and relinquishment;
(n) suspension and modification of the Management Scheme, or cancellation of the Management Scheme, exploration or development permit, and the imposition of monetary penalties, in accordance with Article 51;
(o) procedures for agreed modifications;
(p) enforcement of the Management Scheme;
(q) applicable law to the extent necessary;
(r) effective additional procedures for the settlement of disputes;
(s) provisions to avoid and to resolve conflict with other legitimate uses of Antarctica;
(t) data and information collection, reporting and notification requirements;
(u) confidentiality; and
(v) removal of installations and equipment, as well as site rehabilitation.

Article 48. Approval of the Management Scheme

A Management Scheme prepared in accordance with Articles 45, 46 and 47 shall be subject to approval pursuant to Article 32. Such approval shall constitute authorisation for the issue without delay of an exploration permit by the Regulatory Committee. The exploration permit shall accord exclusive rights to the Operator to explore and, subject to Articles 53 and 54, to develop the mineral resource or resources which are the subject of the Management Scheme exclusively in accordance with the terms and conditions of the Management Scheme.

Article 49. Review

1 Any member of the Commission, or any member of a Regulatory Committee, may within one month of a decision by that Regulatory Committee to approve a Management Scheme or issue a development permit, request that the Commission be convened in accordance with Article 19(2)(b) or (c), as the case may be, to review the decision of the Regulatory Committee for consistency with the decision taken by the Commission to identify the area pursuant to Article 41 and any measures in effect relevant to that decision.

2 The Commission shall complete its consideration within three months of

a request made pursuant to paragraph 1 above. In performing its functions the Commission shall not assume the functions of the Regulatory Committee, nor shall it substitute its discretion for that of the Regulatory Committee.

3   Should the Commission determine that a decision to approve a Management Scheme or issue a development permit is inconsistent with the decision taken by the Commission to identify the area pursuant to Article 41 and any measures in effect relevant to that decision, it may request that Regulatory Committee to reconsider its decision.

Article 50. Rights of Authorised Operators

1   No Management Scheme shall be suspended or modified and no Management Scheme, exploration or development permit shall be cancelled without the consent of the Sponsoring State except pursuant to Article 51, or Article 54 or the Management Scheme itself.

2   Each Operator authorised to conduct activities pursuant to a Management Scheme shall exercise its rights with due regard to the rights of other Operators undertaking exploration or development in the same identified area.

Article 51. Suspension, Modification or Cancellation of the Management Scheme and Monetary Penalties

1   If a Regulatory Committee determines that exploration or development authorised pursuant to a Management Scheme has resulted or is about to result in impacts on the Antarctic environment or dependent or associated ecosystems beyond those judged acceptable pursuant to this Convention, it shall suspend the relevant activities and as soon as possible modify the Management Scheme so as to avoid such impacts. If such impacts cannot be avoided by the modification of the Management Scheme, the Regulatory Committee shall suspend it, or cancel it and the exploration or development permit.

2   In performing its functions under paragraph 1 above a Regulatory Committee shall, unless emergency action is required, seek and take into account the views of the Advisory Committee.

3   If a Regulatory Committee determines that an Operator has failed to comply with this Convention or with measures in effect pursuant to it or a Management Scheme applicable to that Operator, the Regulatory Committee may do all or any of the following:

   (a) modify the Management Scheme;
   (b) suspend the Management Scheme;
   (c) cancel the Management Scheme and the exploration or development permit; and
   (d) impose a monetary penalty.

4 Sanctions determined pursuant to paragraph 3(a) to (d) above shall be proportionate to the seriousness of the failure to comply.

5 A Regulatory Committee shall cancel a Management Scheme and the exploration or development permit if an Operator ceases to have a substantial and genuine link with the Sponsoring State as defined in Article 1(12).

6 The Commission shall adopt general measures, which may include mitigation, relating to action by Regulatory Committees pursuant to paragraphs 1 and 3 above and, as appropriate, to the consequences of such action. No application pursuant to Article 44 may be lodged until such measures have come into effect.

Article 52. Monitoring in Relation to Management Schemes

1 Each Regulatory Committee shall monitor the compliance of Operators with Management Schemes within its area of competence.

2 Each Regulatory Committee, taking into account the advice of the Advisory Committee, shall monitor and assess the effects on the Antarctic environment and on dependent and on associated ecosystems of Antarctic mineral resource activities within its area of competence, particularly by reference to key environmental parameters and ecosystem components.

3 Each Regulatory Committee shall, as appropriate, inform the Commission and the Advisory Committee in a timely fashion of monitoring under this Article.

CHAPTER V. DEVELOPMENT

Article 53. Application for a Development Permit

1 At any time during the period in which an approved Management Scheme and exploration permit are in force for an Operator, the Sponsoring State may, on behalf of that Operator, lodge with the Regulatory Committee an application for a development permit.

2 An application shall be accompanied by the fees established by the Regulatory Committee in accordance with Article 43(2)(b) and shall contain:

   (a) an updated description of the planned development identifying any modifications proposed to the approved Management Scheme and any additional measures to be taken, consequent upon such modifications, to ensure consistency with this Convention, including any measures in effect pursuant thereto and the general requirements referred to in Article 43(3);

   (b) a detailed assessment of the environmental and other impacts of the planned development, taking into account Articles 15 and 26(4);

   (c) a recertification by the Sponsoring State of the technical competence and financial capacity of the Operator and that the Operator has a substantial and genuine link with it as defined in Article 1(12);

(d) a recertification by the Sponsoring State of the capacity of the Operator to comply with the general requirements referred to in Article 43(3);

(e) updated information in relation to all other matters specified in Article 44(2); and

(f) such further information as may be required by the Regulatory Committee or in measures adopted by the Commission.

Article 54. Examination of Applications and Issue of Development Permits

1 The Regulatory Committee shall meet as soon as possible after an application has been lodged pursuant to Article 53.

2 The Regulatory Committee shall determine whether the application contains sufficient or adequate information pursuant to Article 53(2). In performing this function it may at any time seek further information from the Sponsoring State consistent with Article 53(2).

3 The Regulatory Committee shall consider whether:

(a) the application reveals modifications to the planned development previously envisaged;

(b) the planned development would cause previously unforeseen impacts on the Antarctic environment or dependent or associated ecosystems, either as a result of any modifications referred to in subparagraph (a) above or in the light of increased knowledge.

4 The Regulatory Committee shall consider any modifications to the Management Scheme necessary in the light of paragraph 3 above to ensure that the development activities proposed would be undertaken consistently with this Convention as well as measures in effect pursuant thereto and the general requirements referred to in Article 43(3). However, the financial obligations specified in the approved Management Scheme may not be revised without the consent of the sponsoring State, unless provided for in the Management Scheme itself.

5 If the Regulatory Committee in accordance with Article 32 approves modifications under paragraph 4 above, or if it does not consider that such modifications are necessary, the Regulatory Committee shall issue without delay a development permit.

6 In performing its functions under this Article, the Regulatory Committee shall seek and take full account of the views of the Advisory Committee. To that end the Regulatory Committee shall refer to the Advisory Committee all parts of the application which are necessary for it to provide advice pursuant to Article 26, together with any other relevant information.

CHAPTER VI. DISPUTES SETTLEMENT

Article 55. Disputes Between Two or More Parties

Articles 56, 57 and 58 apply to disputes between two or more Parties.

Article 56. Choice of Procedure

1 Each Party, when signing, ratifying, accepting, approving or acceding to this Convention, or at any time thereafter, may choose, by written declaration, one or both of the following means for the settlement of disputes concerning the interpretation or application of this Convention:
    (a) the International Court of Justice;
    (b) the Arbitral Tribunal.
2 A declaration made under paragraph 1 above shall not affect the operation of Article 57(1), (3), (4) and (5).
3 A party that has not made a declaration under paragraph 1 above or in respect of which a declaration is no longer in force shall be deemed to have accepted the competence of the Arbitral Tribunal.
4 If the parties to a dispute have accepted the same means for the settlement of a dispute, the dispute may be submitted only to that procedure, unless the parties otherwise agree.
5 If the parties to a dispute have not accepted the same means for the settlement of a dispute, or if they have both accepted both means, the dispute may be submitted only to the Arbitral Tribunal, unless the parties otherwise agree.
6 A declaration made under paragraph 1 above shall remain in force until it expires in accordance with its terms or until 3 months after written notice of revocation has been deposited with the Depositary.
7 A new declaration, a notice of revocation or the expiry of a declaration shall not in any way affect proceedings pending before the International Court of Justice or the Arbitral Tribunal, unless the parties to the dispute otherwise agree.
8 Declarations and notices referred to in this Article shall be deposited with the Depositary who shall transmit copies thereof to all Parties.

Article 57. Procedure for Dispute Settlement

1 If a dispute arises concerning the interpretation or application of this Convention, the parties to the dispute shall, at the request of any one of them, consult among themselves as soon as possible with a view to having the dispute resolved by negotiation, enquiry, mediation, conciliation, arbitration, judicial settlement or other peaceful means of their choice.
2 If the parties to a dispute concerning the interpretation or application of

this Convention have not agreed on a means for resolving it within 12 months of the request for consultation pursuant to paragraph 1 above, the dispute shall be referred, at the request of any party to the dispute, for settlement in accordance with the procedure determined by the operation of Article 56(4) and (5).

3 If a dispute concerning the interpretation or application of this Convention relates to a measure in effect pursuant to this Convention or a Management Scheme and the parties to such a dispute:

(a) have not agreed on a means for resolving the dispute within 6 months of the request for consultation pursuant to paragraph 1 above, the dispute shall be referred, at the request of any party to the dispute, for discussion in the institution which adopted the instrument in question;

(b) have not agreed on a means for resolving the dispute within 12 months of the request for consultation pursuant to paragraph 1 above, the dispute shall be referred for settlement, at the request of any party to the dispute, to the Arbitral Tribunal.

4 The Arbitral Tribunal shall not be competent to decide or otherwise rule upon any matter within the scope of Article 9. In addition, nothing in this Convention shall be interpreted as conferring competence or jurisdiction on the International Court of Justice or any other tribunal established for the purpose of settling disputes between Parties to decide or otherwise rule upon any matter within the scope of Article 9.

5 The Arbitral Tribunal shall not be competent with regard to the exercise by an institution of its discretionary powers in accordance with this Convention; in no case shall the Arbitral Tribunal substitute its discretion for that of an institution. In addition, nothing in this Convention shall be interpreted as conferring competence or jurisdiction on the International Court of Justice or any other tribunal established for the purpose of settling disputes between Parties with regard to the exercise by an institution of its discretionary powers or to substitute its discretion for that of an institution.

Article 58. Exclusion of Categories of Disputes

1 Any Party, when signing, ratifying, accepting, approving or acceding to this Convention, or at any time thereafter, may, by written declaration, exclude the operation of Article 57(2) or (3) without its consent with respect to a category or categories of disputes specified in the declaration. Such declaration may not cover disputes concerning the interpretation or application of:

(a) any provision of this Convention or of any measure in effect pursuant to it relating to the protection of the Antarctic environment or dependent or associated ecosystems;

    (b) Article 7(1);
    (c) Article 8;
    (d) Article 12;
    (e) Article 14;
    (f) Article 15; or
    (g) Article 37.

2 Nothing in paragraph 1 above or in any declaration made under it shall affect the operation of Article 57(1), (4) and (5).

3 A declaration made under paragraph 1 above shall remain in force until it expires in accordance with its terms or until 3 months after written notice of revocation has been deposited with the Depositary.

4 A new declaration, a notice of revocation or the expiry of a declaration shall not in any way affect proceedings pending before the International Court of Justice or the Arbitral Tribunal, unless the parties to the dispute otherwise agree.

5 Declarations and notices referred to in this Article shall be deposited with the Depositary who shall transmit copies thereof to all Parties.

6 A Party which, by declaration made under paragraph 1 above, has excluded a specific category or categories of disputes from the operation of Article 57(2) or (3) without its consent shall not be entitled to submit any dispute falling within that category or those categories for settlement pursuant to Article 57(2) or (3), as the case may be, without the consent of the other party or parties to the dispute.

Article 59. Additional Dispute Settlement Procedures

1 The Commission, in conjunction with its responsibilities pursuant to Article 41(1), shall establish additional procedures for third-party settlement, by the Arbitral Tribunal or through other similar procedures, of disputes which may arise if it is alleged that a violation of this Convention has occurred by virtue of:

    (a) a decision to decline a Management Scheme;
    (b) a decision to decline the issue of a development permit; or
    (c) a decision to suspend, modify or cancel a Management Scheme or to impose monetary penalties.

2 Such procedures shall:

    (a) permit, as appropriate, Parties and Operators under their sponsorship, but not both in respect of any particular dispute, to initiate proceedings against a Regulatory Committee;
    (b) require disputes to which they relate to be referred in the first instance to the relevant Regulatory Committee for consideration;
    (c) incorporate the rules in Article 57(4) and (5).

CHAPTER VII. FINAL CLAUSES

Article 60. Signature

This Convention shall be open for signature at Wellington from 25 November 1988 to 25 November 1989 by States which participated in the final session of the Fourth Special Antarctic Treaty Consultative Meeting.

Article 61. Ratification, Acceptance, Approval or Accession

1 This Convention is subject to ratification, acceptance or approval by Signatory States.
2 After 25 November 1989 this Convention shall be open for accession by any State which is a Contracting Party to the Antarctic Treaty.
3 Instruments of ratification, acceptance, approval or accession shall be deposited with the Government of New Zealand, hereby designated as the Depositary.

Article 62. Entry Into Force

1 This Convention shall enter into force on the thirtieth day following the date of deposit of instruments of ratification, acceptance, approval or accession by 16 Antarctic Treaty Consultative Parties which participated as such in the final session of the Fourth Special Antarctic Treaty Consultative Meeting, provided that number includes all the States necessary in order to establish all of the institutions of the Convention in respect of every area of Antarctica, including 5 developing countries and 11 developed countries.
2 For each State which, subsequent to the date of entry into force of this Convention, deposits an instrument of ratification, acceptance, approval or accession, the Convention shall enter into force on the thirtieth day following such deposit.

Article 63. Reservations, Declarations and Statements

1 Reservations to this Convention shall not be permitted. This does not preclude a State, when signing, ratifying, accepting, approving or acceding to this Convention, from making declarations or statements, however phrased or named, with a view, *inter alia*, to the harmonisation of its laws and regulations with this Convention, provided that such declarations or statements do not purport to exclude or to modify the legal effect of this Convention in its application to that State.
2 The provisions of this Article are without prejudice to the right to make written declarations in accordance with Article 58.

Article 64. Amendment

1 This Convention shall not be subject to amendment until after the expiry of 10 years from the date of its entry into force. Thereafter, any party may, by written communication addressed to the Depositary, propose a specific amendment to this Convention and request the convening of a meeting to consider such proposed amendment.

2 The Depositary shall circulate such communication to all Parties. If within 12 months of the date of circulation of the communication at least one-third of the Parties reply favourably to the request, the Depositary shall convene the meeting.

3 The adoption of an amendment considered at such a meeting shall require the affirmative votes of two-thirds of the Parties present and voting, including the concurrent votes of the members of the Commission attending the meeting.

4 The adoption of any amendment relating to the Special Meeting of Parties or to the Advisory Committee shall require the affirmative votes of three-quarters of the Parties present and voting, including the concurrent votes of the members of the Commission attending the meeting.

5 An amendment shall enter into force for those Parties having deposited instruments of ratification, acceptance or approval thereof 30 days after the Depositary has received such instruments of ratification, acceptance or approval from all the members of the Commission.

6 Such amendment shall thereafter enter into force for any other Party 30 days after the Depositary has received its instrument of ratification, acceptance or approval thereof.

7 An amendment that has entered into force pursuant to this Article shall be without prejudice to the provisions of any Management Scheme approved before the date on which the amendment entered into force.

Article 65. Withdrawal

1 Any Party may withdraw from this Convention by giving to the Depositary notice in writing of its intention to withdraw. Withdrawal shall take effect two years after the date of receipt of such notice by the Depositary.

2 Any Party which ceases to be a Contracting Party to the Antarctic Treaty shall be deemed to have withdrawn from this Convention on the date that it ceases to be a Contracting Party to the Antarctic Treaty.

3 Where an amendment has entered into force pursuant to Article 64(5), any Party from which no instrument of ratification, acceptance or approval of the amendment has been received by the Depositary within a period of two years from the date of the entry into force of the amendment shall be deemed to have withdrawn from this Convention on the date of the expiration of a further two year period.

4   Subject to paragraphs 5 and 6 below, the rights and obligations of any Operator pursuant to this Convention shall cease at the time its Sponsoring State withdraws or is deemed to have withdrawn from this Convention.

5   Such Sponsoring State shall ensure that the obligations of its Operators have been discharged no later than the date on which its withdrawal takes effect.

6   Withdrawal from this Convention by any Party shall not affect its financial or other obligations under this Convention pending on the date withdrawal takes effect. Any dispute settlement procedure in which that Party is involved and which has seen commenced prior to that date shall continue to its conclusion unless agreed otherwise by the parties to the dispute.

Article 66. Notifications by the Depositary

The Depositary shall notify all Contracting Parties to the Antarctic Treaty of the following:

(a)   signatures of this Convention and the deposit of instruments of ratification, acceptance, approval or accession;

(b)   the deposit of instruments of ratification, acceptance or approval of any amendment adopted pursuant to Article 64;

(c)   the date of entry into force of this Convention and of any amendment thereto;

(d)   the deposit of declarations and notices pursuant to Articles 56 and 58;

(e)   notifications pursuant to Article 18; and

(f)   the withdrawal of a Party pursuant to Article 65.

Article 67. Authentic Texts, Certified Copies and Registration with the United Nations

1   This Convention of which the Chinese, English, French, Russian and Spanish texts are equally authentic shall be deposited with the Government of New Zealand which shall transmit duly certified copies thereof to all Signatory and Acceding States.

2   The Depositary shall also transmit duly certified copies to all Signatory and Acceding States of the text of this Convention in any additional language of a Signatory or Acceding State which submits such text to the Depositary.

3   This Convention shall be registered by the Depositary pursuant to Article 102 of the Charter of the United Nations.

Done at Wellington this second day of June 1988. In Witness whereof, the undersigned, duly authorised, have signed this Convention.

ANNEX FOR AN ARBITRAL TRIBUNAL

Article 1

The Arbitral Tribunal shall be constituted and shall function in accordance with this Convention, including this Annex.

Article 2

1 Each Party shall be entitled to designate up to three Arbitrators, at least one of whom shall be designated within three months of the entry into force of this Convention for that Party. Each Arbitrator shall be experienced in Antarctic affairs, with knowledge of international law and enjoying the highest reputation for fairness, competence and integrity. The names of the persons so designated shall constitute the list of Arbitrators. Each Party shall at all times maintain the name of at least one Arbitrator on the list.

2 Subject to paragraph 3 below, an Arbitrator designated by a Party shall remain on the list for a period of five years and shall be eligible for redesignation by that Party for additional five year periods.

3 An Arbitrator may by notice given to the Party which designated that person withdraw his name from the list. If an Arbitrator dies or gives notice of withdrawal of his name from the list or if a Party for any reason withdraws from the list the name of an Arbitrator designated by it, the Party which designated the Arbitrator in question shall notify the Executive Secretary promptly. An Arbitrator whose name is withdrawn from the list shall continue to serve on any Arbitral Tribunal to which that Arbitrator has been appointed until the completion of proceedings before that Arbitral Tribunal.

4 The Executive Secretary shall ensure that an up-to-date list is maintained of the Arbitrators designated pursuant to this Article.

Article 3

1 The Arbitral Tribunal shall be composed of three Arbitrators who shall be appointed as follows:

    (a) The party to the dispute commencing the proceedings shall appoint one Arbitrator, who may be its national, from the list referred to in Article 2 of this Annex. This appointment shall be included in the notification referred to in Article 4 of this annex.

    (b) Within 40 days of the receipt of that notification, the other party to the dispute shall appoint the second Arbitrator, who may be its national, from the list referred to in Article 2 of this Annex.

    (c) Within 60 days of the appointment of the second Arbitrator, the parties to the dispute shall appoint by agreement the third Arbitrator from the list referred to in Article 2 of this Annex. The third

Arbitrator shall not be either a national of, or a person designated by, a party to the dispute, or of the same nationality as either of the first two Arbitrators. The third Arbitrator shall be the Chairman of the Arbitral Tribunal.

(d) If the second Arbitrator has not been appointed within the prescribed period, or if the parties to the dispute have not reached agreement within the prescribed period on the appointment of the third Arbitrator, the Arbitrator or Arbitrators shall be appointed, at the request of any party to the dispute and within 30 days of the receipt of such request, by the President of the International Court of Justice from the list referred to in Article 2 of this Annex and subject to the conditions prescribed in subparagraphs (b) and (c) above. In performing the functions accorded him in this subparagraph, the President of the Court shall consult the parties to the dispute and the Chairman of the Commission.

(e) If the President of the International Court of Justice is unable to perform the functions accorded him in subparagraph (d) above or is a national of a party to the dispute, the functions shall be performed by the Vice-President of the Court, except that if the Vice-President is unable to perform the functions or is a national of a party to the dispute the functions shall be performed by the next most senior member of the Court who is available and is not a national of a party to the dispute.

2 Any vacancy shall be filled in the manner prescribed for the initial appointment.

3 In disputes involving more than two Parties, those Parties having the same interest shall appoint one Arbitrator by agreement within the period specified in paragraph 1(b) above.

Article 4

The party to the dispute commencing proceedings shall so notify the other party or parties to the dispute and the Executive Secretary in writing. Such notification shall include a statement of the claim and the grounds on which it is based. The notification shall be transmitted by the Executive Secretary to all Parties.

Article 5

1 Unless the parties to the dispute agree otherwise, arbitration shall take place at the headquarters of the Commission, where the records of the Arbitral Tribunal shall be kept. The Arbitral Tribunal shall adopt its own rules of procedure. Such rules shall ensure that each party to the dispute has a full opportunity to be heard and to present its case and shall also ensure that

the proceedings are conducted expeditiously.

2 The Arbitral Tribunal may hear and decide counter-claims arising out of the dispute.

Article 6

1 The Arbiral Tribunal, where it considers that prima facie it has jurisdiction under this Convention, may:
  (a) at the request of any party to a dispute, indicate such provisional measures as it considers necessary to preserve the respective rights of the parties to the dispute;
  (b) prescribe any provisional measures which it considers appropriate under the circumstances to prevent serious harm to the Antarctic environment or dependent or associated ecosystems.

2 The parties to a dispute shall comply promptly with any provisional measures prescribed under paragraph 1(b) above pending an award under Article 9 of this Annex.

3 Notwithstanding Article 57(1), (2) and (3) of this Convention, a party to any dispute that may arise falling within the categories specified in Article 58(1)(a) to (g) of this Convention may at any time, by notification to the other party or parties to the dispute and to the Executive Secretary in accordance with Article 4 of this Annex, request that the Arbitral Tribunal be constituted as a matter of exceptional urgency to indicate or prescribe emergency provisional measures in accordance with this Article. In such case, the Arbitral Tribunal shall be constituted as soon as possible in accordance with Article 3 of this Annex, except that the time periods in Article 3(1)(b), (c) and (d) shall be reduced to 14 days in each case. The Arbitral Tribunal shall decide upon the request for emergency provisional measures within two months of the appointment of its Chairman.

4 Following a decision by the Arbitral Tribunal upon a request for emergency provisional measures in accordance with paragraph 3 above, settlement of the dispute shall proceed in accordance with Articles 56 and 57 of this Convention.

Article 7

Any Party which believes it has a legal interest, whether general or individual, which may be substantially affected by the award of an Arbitral Tribunal, may, unless the Arbitral Tribunal decides otherwise, intervene in the proceedings.

Article 8

The parties to the dispute shall facilitate the work of the Arbitral Tribunal and, in particular, in accordance with their law and using all means at their disposal, shall provide it with all relevant documents and information, and enable it, when necessary, to call witnesses or experts and receive their evidence.

Article 9

If one of the parties to the dispute does not appear before the Arbitral Tribunal or fails to defend its case, any other party to the dispute may request the Arbitral Tribunal to continue the proceedings and make its award.

Article 10

1 The Arbitral Tribunal shall decide, on the basis of this Convention and other rules of law not incompatible with it, such disputes as are submitted to it.

2 The Arbitral Tribunal may decide, ex aequo et bono, a dispute submitted to it, if the parties to the dispute so agree.

Article 11

1 Before making its award, the Arbitral Tribunal shall satisfy itself that it has competence in respect of the dispute and that the claim or counterclaim is well founded in fact and law.

2 The award shall be accompanied by a statement of reasons for the decision and shall be communicated to the Executive Secretary who shall transmit it to all Parties.

3 The award shall be final and binding on the parties to the dispute and on any Party which intervened in the proceedings and shall be complied with without delay. The Arbitral Tribunal shall interpret the award at the request of a party to the dispute or of any intervening Party.

4 The award shall have no binding force except in respect of that particular case.

5 Unless the Arbitral Tribunal decides otherwise, the expenses of the Arbitral Tribunal, including the remuneration of the Arbitrators, shall be borne by the parties to the dispute in equal shares.

Article 12

All decisions of the Arbitral Tribunal, including those referred to in Articles 5, 6 and 11 of this Annex, shall be made by a majority of the Arbitrators who may not abstain from voting.

Index

Agreed Measures for the
    Conservation of Antarctic Fauna
    and Flora (AMCAFF), 3, 4, 75
Alaska, 38
Amundsen Sea, 15
Andes, 33, 34, 36
Antarctic and Southern Ocean
    Coalition (ASOC), 68–9, 73, 77
Antarctic bases *see* scientific stations
Antarctic Bottom Water, 11, 20
Antarctic Circumpolar Current, 10
Antarctic continent geology 14–15,
    30–2
Antarctic Convergence, 3, 16
Antarctic environment, 4–7, 8–10,
    18–21, 44, 60–3, 69–72
Antarctic Minerals Act, 87
Antarctic Peninsula, 15, 25, 30, 33,
    37, 61
Antarctic science, 8–24
Antarctic Treaty, 2–3, 5–6, 39, 48–
    9, 59, 68, 74–5, 81–3
    Contracting Parties, 116–17
    prohibitions, 45, 46, 50, 63
    text of, 110–16
Antarctic Treaty Consultative
    Party, 2, 44, 55, 96, 99, 100
    meetings of, 3, 4, 45, 53, 57, 58–9,

62, 73, 76, 83, 88, 95, 98, 99, 100,
    102
Antarctic Treaty system, 5, 20–1,
    50–2, 56, 64, 78, 81, 89, 95–6,
    102
Antarctica
    and global climate, 5, 9–10, 11–
    13
    exploration of, 1
antimony, 36
Arctic, 11, 38–9, 97
Argentina, 40, 71, 81
Asia, 36
Australia, 5–7, 27, 30, 36, 37, 53–65,
    73, 88, 98, 99, 100
Australian Conservation
    Foundation, 54
Australian Institute of International
    Affairs, 46

banded iron formations *see* iron
Beaufort Sea, 39
Beeby, Christopher, 81
Belgium, 55, 87, 99
Biological Investigations of Marine
    Antarctic Systems and Stocks
    (BIOMASS), 16
biology, 16–18

Bransfield Strait, 37
British Antarctic Survey, 11
British Antarctic Territory
  Ordinance, 87
British Isles, 33

Canada, 39
carbon dioxide *see* 'greenhouse
  effect'
cerium, 34
Chile, 40, 57, 58, 59, 74, 81, 88, 89,
  98
chlorofluorocarbons (CFCs), 11–12,
  70 *see also* greenhouse effect
chromium, 33, 34, 36
climate, 9–10 *see also* 'greenhouse
  effect'
climatology, 10–14
coal, 27, 32, 38
Coats Land, 36
cobalt, 34, 36
'Common Heritage' concept, 20–1
comprehensive environment
  protection convention (*see also*
  World Park), 6–7, 45, 53–67, 74,
  88, 97–100
continental reconstructions *see*
  Gondwana
Convention on the Conservation of
  Antarctic Marine Living
  Resources (CCAMLR), 3, 4, 76,
  83, 84–5
  Commission for, 3, 84
  Scientific Committee, 85
Convention for the Conservation of
  Antarctic Seals (CCAS), 3, 76,
  83–4
Convention on the Regulation of
  Antarctic Mineral Resource
  Activities (CRAMRA), 3–4, 6–7,
  25, 39–40, 44–52, 57, 60–5, 77–9,
  83, 85–7, 96–103

Commission for, 4, 47, 78, 86, 90
  prospecting under, 46–8
  signatories, 51
  text of, 118–167
Cook, Captain James, 1
copper, 33, 34, 36
Cousteau Foundation, 54

DDT, 70
Drake Passage, 15
Dronning Maud Land, 36
Dufek Massif, 34
Dumont d'Urville airstrip, 75

East Antarctica, 30, 37
economic geology, 26, 38
economic mineral deposits, 27
environmental damage, 39, 61–2,
  68, 78, 96
  definition of, 78
environmentalist's view, 68–80
Europe, 36
European Parliament, 55

Falkland Islands, 17
fish, 2, 9, 17, 20, 69
fossils, 15
France, 6–7, 53–5, 57, 65, 74, 88,
  98, 99, 100

gas *see* hydrocarbons
geological mapping, 25, 26, 29, 37
geology and geophysics, 14–15, 26,
  48
geospace, 13–14, 20
Germany, Federal Republic of, 55,
  99
global warming *see* 'greenhouse
  effect'
gold, 36
Gondwana, 10, 14, 15, 20, 34–7
Greater Antarctica, 14, 15

'greenhouse effect', 5, 11–13, 19, 68, 70
Greenpeace, 54, 68, 71, 75, 79
Gulf of Mexico, 39

habitat destruction, 70
Hawke, R. J. L., 7n, 65n
  foreword by, vi–vii
hydrocarbons, 9, 14, 19, 26, 27, 29, 31, 32, 37, 38, 61

icebergs, 62
icesheet, 8, 10, 12–13, 15, 16, 19, 20, 25, 29
India, 99
India mineral province, 36
inland waters, 18–19, 69
International Council of Scientific Unions (ICSU), 10, 19
International Geophysical Year, 2
International Geosphere Biosphere Programme (IGBP), 19
international law, 46, 49
International Polar Year, 1, 2
international relations, 50–2 *see also* political factors
International Whaling Commission, 2
ionosphere *see* geospace
iron, 33, 34, 36
Italy, 55, 99

Kenneally, Thomas, 1
krill, 2, 9, 16–17, 20, 69, 72

lanthanum, 34
Larsen Basin, 37
lead, 3, 36, 38
legal jurisdiction, 81, 82, 85–6, 87, 90–1, 101
  see also territorial sovereignty
Lesser Antarctica, 14, 15

Liability Protocol, 87, 90, 98–9, 102
  see also CRAMRA

magnetosphere *see* geospace
manganese, 33, 36
marine ecosystem, 3, 9, 69
McMurdo Station *see* scientific stations
meteorology, 10–14
methane *see* 'greenhouse effect'
mineral exploration and exploitation, 5–7, 9, 20, 45, 62, 72
  moratorium on, 49, 96, 101, 102
mineral resources, 9, 14, 19, 25–43, 97, 101
  definitions of, 26–9
  see also individual mineral entries
mineralisation *see* mineral resources
Minerals Convention *see* Convention on the Regulation of Antarctic Mineral Resource Activities
mining *see* mineral exploration and exploitation
molybdenum, 33, 34, 36

National Science Foundation (US), 71
'natural reserve – land of science,' 6, 98
New Zealand, 14, 55, 57, 58, 59, 65, 74, 88–9, 91, 98
nickel, 34, 36
niobium, 34
nitrous oxide *see* 'greenhouse effect'
North America, 27, 36

oceans and ocean currents, 10–11
oil *see* hydrocarbons
overfishing, 9, 17, 72, 76
ozone layer, 5, 11–12, 19, 68

palaeobotanic studies, 15
palaeomagnetic studies, 15, 30
Palmer Station *see* scientific stations
plankton, 12
plate tectonics, 14–15, 20, 34
platinum, 34, 36
polar front *see* Antarctic
convergence
polar vortex, 11
political factors, 39–40, 44–52, 54–6, 100
pollution, 20
Prince Charles Mountains, 32
prospecting, definition of, 47
protocol on liability *see* liability
protocol

Rocard, M., 6, 7n, 54
Ross Sea, 15, 32, 37, 61

Schlesinger, Arthur M., Jr., 50
Scientific Committee on Antarctic
Research (SCAR), 10, 16, 19, 61
scientific stations, 2, 8, 18, 62, 71,
102
list of, 106–8
location map, 109
Scotia Sea, 15
sea birds, 16, 17, 69
sea level, 13 *see also* 'greenhouse
effect'
seals, 2, 9, 16–17, 20, 69, 76 *see also*
Convention for the
Conservation of Antarctic Seals
seismic surveys, 29, 31–2
Seymour Island, 15
shipping accidents, 62, 71, 72, 95,
97
silver, 36, 38
solar wind, 13
soils, 8, 62, 71
South Africa, 27, 36

South America, 14, 15, 37
South Pole, 1
Southern Ocean, 9, 12, 20, 61
sovereignty *see* territorial
sovereignty
Specially Protected Areas, 39, 57–8,
83, 95
squid, 9, 17, 20
Sweden, 55, 57, 58, 88, 90, 98

Terre Adélie, 36
terrestrial ecosystem, 18, 69
territorial claims, 2, 40, 81, 87
map of, 105
territorial sovereignty, 40, 77, 81–2,
87, 96, 101
thorium, 34
tin, 34, 36
titanium, 34
tourist activities, 98, 101
Transantarctic Mountains, 15, 30,
32, 36
tungsten, 36

ultra-violet radiation, 12, 13
Union of Soviet Socialist Republics,
38, 39, 76, 81, 99
United Kingdom, 40, 54, 55, 59, 81,
82, 87
United Nations, 85, 102
trusteeship, 20, 102
General Assembly, 65
United States of America, 38, 39,
54, 56, 58, 63, 74, 77, 81, 88, 89,
98, 99
uranium, 34, 36

vanadium, 34

waste disposal, 20, 71
Code of Conduct, 76, 83, 95
Watts, Sir Arthur, 77

Weddell Sea, 15, 32, 37, 61
West Antarctica, 30, 36, 37
whales, 2, 9, 16, 17, 20, 69
Wilkes land, 36
World Park, 20, 53, 68–9, 73, 77,
  97–8 *see also* comprehensive
  environment protection
  convention.

Yilgarn mineral province, 36

zinc, 33, 36, 38
zirconium, 34